EcoProduction

Environmental Issues in Logistics and Manufacturing

Series Editor

Paulina Golinska, Poznan, Poland

For further volumes:
http://www.springer.com/series/10152

About the Series

The EcoProduction Series is a forum for presenting emerging environmental issues in Logistics and Manufacturing. Its main objective is a multidisciplinary approach to link the scientific activities in various manufacturing and logistics fields with the sustainability research. It encompasses topical monographs and selected conference proceedings, authored or edited by leading experts as well as by promising young scientists. The Series aims to provide the impulse for new ideas by reporting on the state-of-the-art and motivating for the future development of sustainable manufacturing systems, environmentally conscious operations management and reverse or closed loop logistics.

It aims to bring together academic, industry and government personnel from various countries to present and discuss the challenges for implementation of sustainable policy in the field of production and logistics.

Paulina Golinska
Editor

Environmental Issues in Automotive Industry

Editor
Paulina Golinska
Poznan University of Technology
Poznan
Poland

ISSN 2193-4614 ISSN 2193-4622 (electronic)
ISBN 978-3-642-23836-9 ISBN 978-3-642-23837-6 (eBook)
DOI 10.1007/978-3-642-23837-6
Springer Heidelberg New York Dordrecht London

Library of Congress Control Number: 2013942155

Printed on acid-free paper

Springer is part of Springer Science+Business Media (www.springer.com)

Preface

The automotive industry is a sector where environmental impact must be taken into consideration in many ways. First the production processes need to be less harmful for the environment. Then the product itself must be optimized for Middle-of-life and End-of-life phase.

The automotive industry has applied a life cycle approach as one of the major focus. This approach is highlighted in fulfillment of following goals:

- steady improvement in vehicle recovery rates,
- increased use of renewable resources and recycled materials,
- increased utilization of used parts,
- reduction of hazardous substances like lead, mercury, cadmium, and hexavalent chromium,
- reduction of CO_2 emission.

This book entitled "Environmental Issues in Automotive Industry" aims to present the emerging environmental issues in automotive industry. The automotive industry is one of the most environmental aware manufacturing sectors. Product take-back regulations influence design of the vehicles, production technologies and also the configuration of automotive reverse supply chains. The business practice comes every year closer to the closed loop supply chain concept which completely reuses, remanufactures, and recycles all materials.

The book covers the emerging environmental issues in automotive industry through the whole product life cycle. In this book the focus is placed on a multidisciplinary approach. It presents viewpoints of academic and industry personnel on the challenges for implementation of sustainable police in the automotive sector. Authors present in the individual chapters the result of the theoretical and empirical research related to the following topics:

- sustainability in automotive industry,
- tools and methods for greener decision making,
- recovery of end-of-life vehicles.

This book includes research contributions of geographically dispersed authors from Europe, North America, and Asia. It is a clear indication of a growing interest in sustainable development and environmental friendly production and logistics solutions. The high scientific quality of the chapters was assured by a rigorous blind review process implemented by the leading researchers in the field from Canada, Germany, Poland, Spain, and the USA.

This monograph provides a broad scope of current issues important for the development of environmentally friendly management in automotive sector. It is a composition of theoretical trends and practical applications. The advantage of this book is presentation of country-specific applications from number of different countries around world.

I would like to thank all Authors who responded to the call for chapters and submitted manuscripts to this volume. Although not all of the received chapters appear in this book, the efforts spent and the work done for this book are very much appreciated.

I would like to thank all reviewers whose names are not listed in the volume due to the confidentiality of the process. Their voluntary service and comments helped the authors to improve the quality of the manuscripts.

Paulina Golinska

Contents

Part I
Sustainability in Automotive Industry

Environmental Friendly Practices in the Automotive Industry

Paulina Golinska and Monika Kosacka

Abstract The automotive industry is one of the most environmental aware sectors of an economy. Car is a very complex product not only due to thousands of components used in a production process and many people involved in this process but especially for the reason that, it creates threats for the environment at each stage of its life cycle. The aim of this chapter is to provide the review of current environmental friendly practices in the automotive industry regarding reducing strategy, reusing strategy and recycling strategy used by car makers.

Keywords Environment · Reduce · Reuse · Remanufacturing · Recycle

1 Introduction

Environmental focus is one of the leading trends in many industries. In order to preserve the environment for the next generation people should strive to sustainable development in all activities. Therefore it can be concluded that environment became a part of present economy affected each activity, including automotive industry, especially that intensive development of the automotive industry is a source of hazards for the environment. The introduction of EU regulations for this industry like End-Life-Vehicles (ELV Directive 2000/53/EC) and directive on type approval of vehicles for reusability, recyclability, and recoverability (RRR 2005/64/EC) has created a need for the new business practices in area of materials

P. Golinska (✉) · M. Kosacka
Poznan University of Technology, Strzelecka 11 60-965 Poznań, Poland
e-mail: paulina.golinska@put.poznan.pl

M. Kosacka
e-mail: kosacka.monika@gmail.com

P. Golinska (ed.), *Environmental Issues in Automotive Industry*,
EcoProduction, DOI: 10.1007/978-3-642-23837-6_1,

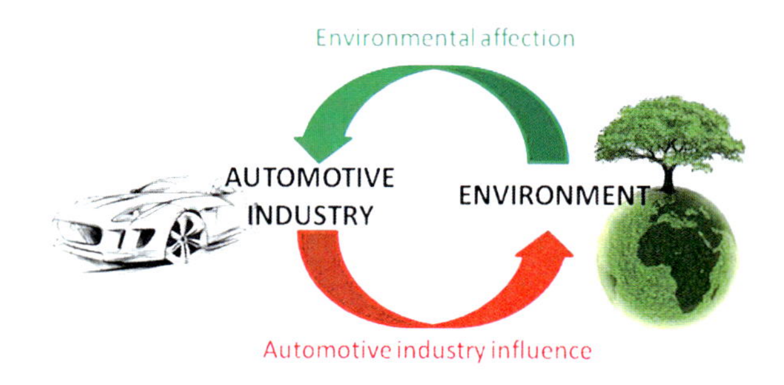

Fig. 1 Feedback between environment and automotive industry

management. Authors define a feedback between an environment and the automotive industry presented at Fig. 1.

There are four factors affecting the automotive industry: technology, market, customers and the most influential factor—environment. Environment is a source of materials, infrastructure necessary for manufacturing but everything is limited. Environmental restrictions have a positive side, because there is a big pressure to avoid wastes during every stages of car life cycle, what let manufacturers to reduce costs and to be more competitive. Not without significance for the environment is that customers present higher level of ecological awareness and there is a trend of buying eco-friendly products.

The process of car or automotive parts manufacturing is very complex, regarding usage of numerous resources and different technologies. It results in potential threats for:

- polluting water, soil and air;
- noise;
- creating waste and landfills;
- damaging land use;
- overutilization of materials;
- disruptions of ecosystem; etc.

Those issues should be perceived in relation to all stages of car life, which are presented at Fig. 2.

Main stages of car life cycle important from environmental point of view are dependent by each other. Beside those relationships there are external factors which have an impact on them. Authors define 4 main stages of vehicle life cycle:

1. Design

There are made crucial decisions about car's construction and production process. In this phase a number of innovations might appear which are making the product more eco-friendly regarding the subsequent life cycle stages.

2. Manufacturing with logistics support

This is the implementation of the project. This stage is characterized by huge complexity, big number of operations and parts, big diversity. From environmental

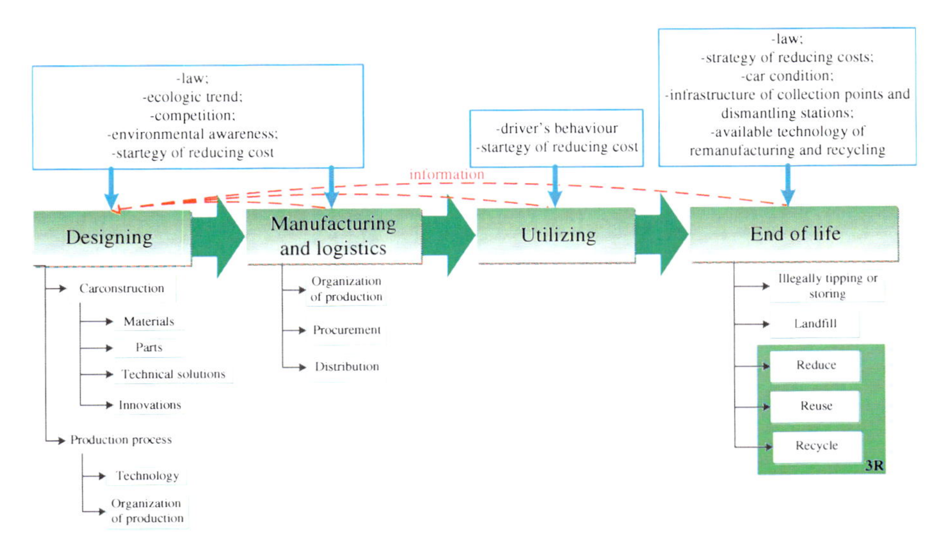

Fig. 2 Car life cycle based on (Parkinson and Thompson 2003)

perspective it might create some threats to the environment mainly in the welding and painting processes. Moreover most of the automotive manufacturers applying just-in-time and just-in-sequence strategy for their components deliveries. The big scope of transport operations is a main environmental burden coming from final assembly.

3. Utilization

This is the longest phase of life of a car. It covers all issues related to the after-sale usage of each vehicle. The environmental burdens which appear in this phase are connected with normal "consumption" of product and its maintenances. When the need for repairs appears, there are necessary: new spare parts, remanufactured components or used parts from recycling. Moreover during usage phase a number of emissions appear which might be potentially dangerous for the environment.

4. End of life

In this stage three different options might be implemented: landfill, illegally tipping and storing of a vehicle or revalorization. This stage is referred to managing the stream of used cars and their components. At this stage are distinguished 3 main strategies used by car manufacturers (described as 3R) (Parkinson and Thompson 2003):

1. Reduce strategy—creating solutions to reduce waste and as a result of it increasing recycling opportunities through the all process of cars' manufacturing.

2. Reuse strategy—continuing to use an item after it has been relinquished by its previous user, rather than destroying, or recycling it. The extreme case of this strategy is reuse "as is" which refers to the reuse of a product with minimal reprocessing.
3. Recycle strategy related to: creating new consumer goods or new car parts from materials obtained during dismantling process of old car or producing new parts from recycled consumer products (recyclable and non-recyclable materials).

In European conditions the three strategies were addressed in the RRR Directive 2005/64/EC on type approval of vehicles for reusability, recyclability, and recoverability. This Directive came into force in December 2005 and requires cars and light vans (M1/N1), newly introduced to the market after December 2008 to be 85 % reusable and/or recyclable and 95 % reusable/recoverable by mass (Directive 2005/64/EC). Focus on recyclability has driven the new model planning process. Newly applied advanced recycling methods (post shredder treatment) allow nowadays the recycling and recovery of literally all materials. Moreover there is a shift in design approach so called product modularity. It allows improving disassembly operations. To speed up the dismantling operations all components are labeled in accordance with international ISO standards, enabling materials to be sorted according to their type. In order to reach the challenging goal of 95 % recovery target by 2015 some efficient material separation technologies for end-of-life vehicles are promoted that allow the utilization for shredder residue and boosting the usage of recycled materials for some specific car components. The Japan Automotive manufacturers are also obliged to reach the goal of recycling rate by over 95 % till 2015. US manufacturers don't face as strict regulation as in EU or Japan. On average they reach the goal of 75 % materials recoverability and recycling ability.

2 Reducing Strategy

"Reduce" is the key word of this strategy. Figure 3 presents the mind map of issues related to reducing strategy in the automotive industry. There are considered four areas of adopting reducing strategy by carmakers (FIAT 2011; Toyota 2011):

1. Logistics activities

This area includes a range of all activities required to handle materials, components and products across the supply chain, from suppliers to manufacturers and final customer. Reducing strategy in the transport is driven by a series of actions, ranging from the reducing emissions linked to the transportation of finished goods particularly by using alternatives means of transport to road (rail, sea) and optimization of transport capacity in order to achieve reduction of CO_2 emission. The second point is packaging with the particular focus on minimizing packages and

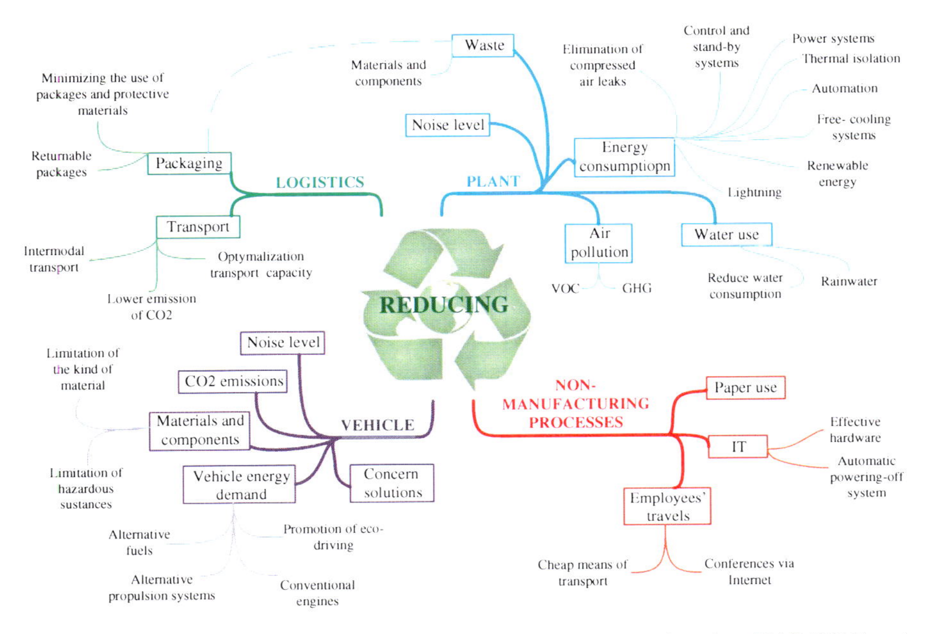

Fig. 3 The mind map of reducing strategy in the automotive industry based on FIAT (2011) and Toyota (2011)

protective materials and increasing the use of reusable packages (for example containers).

2. Vehicle utilization

At the design stage, there are made crucial decisions about the environmental impact of a car during its utilization, including: vehicle energy demand, the noise level, emissions of CO_2, used materials and components, other concern solutions of carmakers. The greatest attention is related to the following issues:

- Optimizing the ecological performance of conventional engines.
- Alternative fuels such as: natural gas, biofuels, biomethane.
- Alternative propulsion systems including: conventional hybrids, plug-in hybrids, fully electrified and range-extended electric vehicles.
- Promotion of eco-driving among consumers.

Decisions about car energy consumption affect the noise level and greenhouse gases (GHG) emissions, moreover there are installed some systems cutting gases' emissions. Steps taken for used materials aimed at limitation kinds of materials and components used in manufacturing process and limiting the use of potentially hazardous substances with promoting their substitution wherever it is possible. Beside those solutions there are introduced many innovations, for example Fiat contributes the following solutions reducing fuel consumption and emissions (FIAT 2011):

- Start&Stop technology (shutting off the engine at the stoppage idling and restarting after engaging the clutch),
- Gear Shift Indicator (virtual co-pilot suggesting gears shifts to reduce emissions and energy demand),
- Smart Alternator (system enabling battery charging independently).

Automakers try to apply reduce strategy also for the activities unrelated, which are not directly related to the manufacturing process. Some examples of such actions which can be found in automotive sector are:

- Rationalizing employ travels by participating in conferences via the Internet, giving priority to cheaper means of transport such as rail.
- Changes in the IT: effective hardware, reduction in PC power consumption by automatically powering down PC when it is not in use.
- Reduction is paper usage by application of the IT systems to communicate downstream and upstream of the supply chain.

3 Reusing Strategy

Automotive industry is a part of industrial sector that consumes large volumes of material and energy, therefore it is essential to save them, what is the result not only of financial factors but also ecological activity. Possibilities of reusing are defined by designers and engineers at the beginning of car life cycle when materials and components are selected to use. Reusing is a way to stop waste at the source.

3.1 Reusing Strategy Possibilities

There is some confusion about regarding the classification of the reuse options. In the rather narrow way the reuse variants can be defined as:

Reuse can be divided into (www.remanufacturing.org.uk):

- straight reuse by other user with lower quality expectations (less developed markets),
- refurbishment—cleaning, lubricating or other improvement,
- repair/rebuilt
- redeployment & cannibalization—using working parts elsewhere.

Rebuilt operations are applied for automotive industry mainly for heavy duty equipment such as bus and truck fleets, farming equipment and construction equipment when new parts many be difficult or impossible to obtain. In this paper we will extend this classification and treat remanufacturing as a part of reuse strategy. Remanufacturing is the only option that requires a full treatment process to guarantee the performance of the finished object ().

It is important to distinguished reuse from recycle strategy. Recycling activities require the destruction of the product to its components and materials so they can be melted, smelted or reprocessed into new forms ().

3. Plant functioning

With the functioning of the plant are related many dangers for the environment what is a result of manufacturing process and all activities related to them. Reducing strategy at this field focuses on: energy consumption and air pollution by GHG, water use, waste management and noise level. Energy is the most important store for a car plant which is very energy-intensive, what is associated to huge amount of greenhouse gases (GHG). Car makers put a lot of efforts to reduce energy demand and GHG emissions such as (FIAT 2011):

- renewable energy (solar energy, wind energy, hydro energy, paint fumes or landfill gases as a energy source);
- high-efficiency motors and electric motor inverters in the area of power;
- high-efficiency lighting systems;
- frequency control and smart stand-by systems for equipment;
- thermal isolation of facilities;
- elimination of compressed air leaks;
- automated control systems increasing energy efficiency;
- free-cooling systems rather than electric air-conditioners.

A big problem of car manufacturing are volatile organic compounds (VOCs) emissions from painting operations which are reduced by innovation in paint application and materials technology. Ford uses VOC to produce energy. Noise level is monitored and reduced where it is possible by the installation of sound-absorbing elements and noise abatement walls, for example.

Car manufacturers take actions to implement potential solutions for reducing overall water consumption and ensuring the high quality of discharged water by minimizing emissions of hazardous substances to water from manufacturing. The common practice is using rainwater.

Beside the management of water and energy car makers manage waste seeking which is reflected in the financial results. Waste management is related to packages (discussed at the Logistics area) and materials, components used at manufacturing process. In the automotive industry landfills are the extremity—when generated waste is not able to be reused or recycled, it is disposed of, seeking to use technologies with minimal environmental impact. More often car makers strive to use materials and components recyclable, susceptible to re-use because it is connected to financial profits, what will be described in more detail in the next sections.

4. Non-manufacturing processes

Recycling is putting used materials back into the manufacturing chain at a very basic level. Reusing means continuing using of a product, not destroying it what is a main characteristic in recycling (Parkinson and Thompson 2003). Reusing is used to indicate all forms of material product and component reutilization (Parkinson and Thompson 2003).

There are some difficulties with determining the meanings' relationships between all terms related to reusing strategy. In Fig. 4 authors present the scheme of possibilities of reusing strategy. It is a modification of division founded by Parkinson and Thompson (2003).

Following Parkinson and Thompson (2003) distinguish three types of activities as components of reusing strategy: further use, reuse "as is" and product reprocessing.

Further use. Used products are utilized for a different purpose than it was originally intended. Example: using tires on harbor walls to cushion vessels.

Reuse "as is". Used products are utilized at the same purpose they were designed for through minimal or no reprocessing. The minimum contribution to restore parts to working conditions includes for example: cleaning, cursory inspection, etc. Example: obtaining mirrors from a used vehicle. Lack of precise definitions provide to fuzzy boundaries between terms what results in artificial divisions such as treating reprocessing and reconditioning as two terms independent one from another. Authors presents the statement that product reprocessing and recondition are synonyms.

Product reprocessing consists of all actions involving more than only superficial works, required to return the condition for sale (Parkinson and Thompson 2003). This definition is too "scant" for the authors, because the key for right division is taking into account appropriate criteria. They suggest: warranty and the level of quality of product in comparison to new product as the criteria of distinction. Thanks to that they were able to show how do they depend on each other: repair, remanufacturing and so called "recovery characteristics" (what causes the most problems for authors with this term).

Remanufacturing. Remanufacturing includes disassembling, cleaning, refurbishing, replacing parts (as necessary) and reassembling a product in such a manner that the part is at least as good as, or better than, new in terms of appearance, reliability and performance (Oleszczuk 2003; Parkinson 2001). The

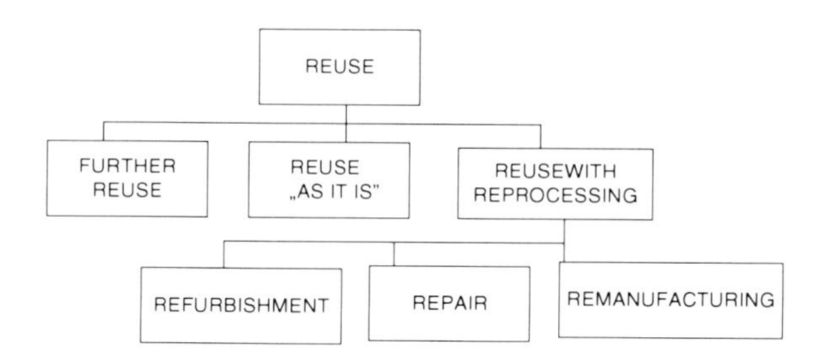

Fig. 4 Possibilities of reusing strategy modified from Parkinson and Thompson (2003)

greatest emphasis of this definition is placed on the condition of remanufactured product—"as good or better than new" with a warranty at least as new item.
Repair is related to actions taken to return a product's functioning condition after detected failure, particularly at the service. Repairs are not equivalent to return its state as good as new. Repairs are related to warranties which are very often shorter than new products (Lund 1983).
Refurbishment is the reprocessing of product in order to provide required functionality. Usually is it connected with some improvement in existing product parameters. For example it might require some improvement of product existing operating software.

3.2 Remanufacturing

Remanufacturing allows companies to capture the residual value-added in forms of materials, energy and labor. Remanufacturing is the most advanced form of reprocessing operations, but in case of End-of-Life products or almost at this stage sometimes it is not appropriate. The reason is the cost and complexity of remanufacturing operations. Lund (1998) has identified seven criteria for profitable remanufacturability:

- the product is a durable good,
- the product fails functionally,
- the product is standardized and the parts are interchangeable,
- the remaining value-added is high,
- the cost to obtain product is low compared to the remaining value,
- the product technology is stable,
- the costumer is aware that remanufactured goods are available.

The remanufacturing industry of the automotive components is large and important due to economic, societal and environmental benefits. Up to two-thirds of remanufacturing businesses globally is estimated to involve auto parts (Steinhilper 1998).

The environmental benefits are related to energy and material savings gained through reusing parts, what is a result of saving the value concluded into the component during manufacturing process. The size of saving is associated to a huge amount of parts used at the automotive industry. According to studies at the automotive industry, approximately 85 % of the energy expended in components' manufacture was preserved at the remanufactured item (Henstock 1988).

Starters and alternators are the most typical products to be remanufactured due to the fact that most car require two of each throughout their lives, and these two components are mass produced and remanufactured by thousands of companies. Alternator and starter motors are ideal candidates for a profitable remanufacturing, as these components have very high production volumes and low un-manufacturability (Severengiz et al. 2008).

To present disparities at material and energy use of manufactured and remanufactured car parts, authors adduce two examples: starters and alternators, which are the most frequently remanufactured car units. Results of those studies are presented in Fig. 5.

According to previous research (e.g. REMAN 2013): manufacturing process of one new starter on average requires more than eleven times the amount of energy of a remanufactured one. In the case of alternators new one requires about seven times more energy than a remanufactured one. In a case of material demand with the same amount of the material it is possible to manufacture one new part or 8 remanufactured alternators, in the case of starters—9 remanufactured starters. This example underline very clearly issue of materials and energy saving during manufacturing.

Remanufacturing process is a sequence of activities required to obtain remanufactured item. Authors present a remanufacturing as a flowchart in Fig. 6.

Remanufacturing process consist of six main activities including (Amezquita et al. 1998; Guide 2000; Parkinson 2001):

(a) Disassembling—products are disassembled to the level of a part. Reusable parts are passed to next operations. Other elements may be recycled or they are disposed.
(b) Cleaning—removing all contamination, including degreasing, derusting, removal coatings of the surface as a paint.
(c) Inspection and sorting—sorting items into groups with assessing the parts' reusability and possibilities of reconditioning.
(d) Reprocessing—includes: milling, turning, grinding, material deposition, heat treatment, welding, powder coating, chroming, painting.

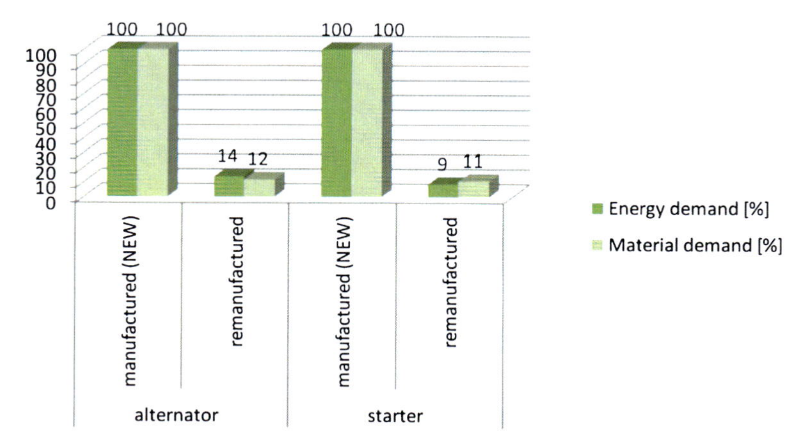

Fig. 5 Remanufacturing and manufacturing demand of energy and material of starters and alternators (REMAN 2013)

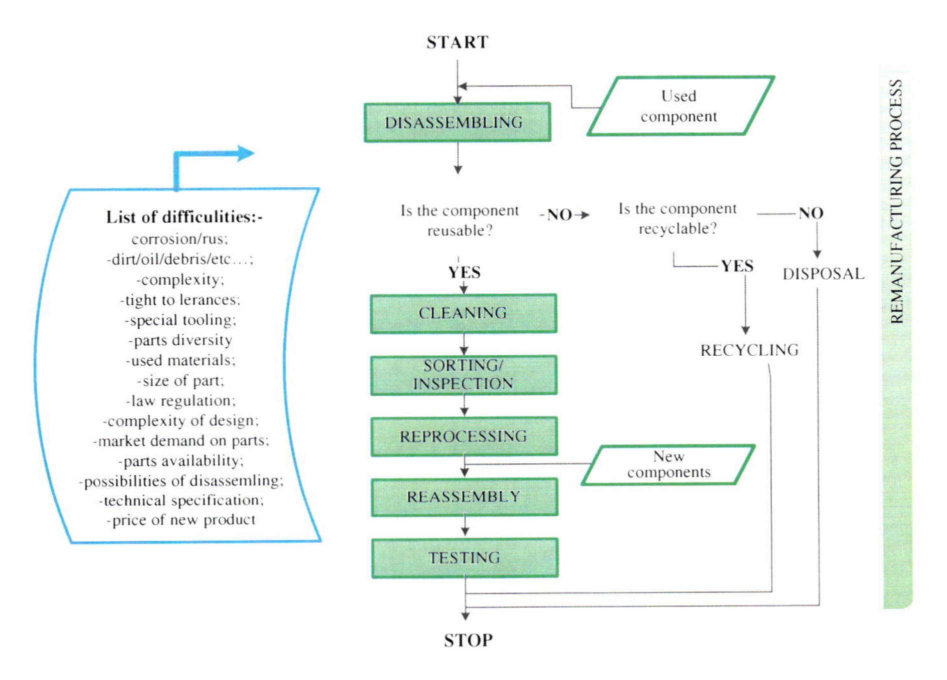

Fig. 6 Remanufacturing process based on Amezquita et al. (1998), Guide (2000) and Parkinson and Thompson (2003)

(e) Reassembling—it is a process of assembling with used components (sometimes with new elements).
(f) Testing—every remanufactured part is tested to preserve efficiency at the level of 100 %. Remanufactured parts obtain a warranty.

Remanufacturing process is associated to many difficulties considered in Fig. 6.

Remanufacturability is the result of decisions at the designing stage of car life cycle. Existing criteria for remanufacturability include (Guide 2000):

1. Durability of a product.
2. Failure product's functions.
3. Product is a standardized good with interchangeable parts.
4. Low cost of obtaining failed product.
5. Stable technology of production.
6. Awareness of costumers of availability of remanufactured products.

Designing for remanufacturing is associated to facilities at the level of a part from the point of view of each activity in remanufacturing process (Amezquita et al. 1995).

Car producers are making some Design-for-Remanufacturing guidelines by setting the criteria that facilitate the remanufacturing of analyzed part.

In Table 1 are presented criteria that facilitate the remanufacturing of a Chrysler LHS door.

The method is based on a method proposed by Pahl and Beitz (1984). The requirements of a product are classified into several main categories which consist of different subcategories. There is presented a specification of the requirements related to the each category. There are: Demands (D)—defined as requirements that must be satisfy regardless of the circumstances and Wishes (W)—defined as requirements that should be taken into consideration whenever it is possible. This example show how remanufacturability can be improved using a list of specified criteria.

Car parts are remanufactured very often, especially due to economic factors. The list of remanufactured car parts is long and includes (APRA 2013):

- Air Brakes
- Air Conditioners
- Alternators
- Brake Calipers
- Carburettors
- Clutches
- CV Drives
- Cylinder Heads
- Differentials
- Electronic Units
- Engines
- Fan Motors
- Heater Blowers
- Water Pumps
- Turbochargers
- Front Axles
- Fuel Pumps
- Fuel Injectors
- Generators
- Master Cylinders
- Pumps
- Power Steering
- Rack and Pinions
- Starters
- Steering Units

The industry leader in terms of remanufacturing Toyota already in 2002, introduced on all European markets, remanufactured compressors for air conditioners and power steering gears. The next step was the introduction at the beginning of 2003 the remanufactured engines with accessories and without fittings and cylinder heads. In addition, Toyota in Europe coordinates the development and implementation of effective systems of reimbursement of exchanged components, which are directed to the main European distribution center, instead of directly from the dealership to the suppliers. Every year an increasing number of the remanufactured parts are being offered. The automaker Volkswagen AG uses worn parts for production "the replacement", having both the value and the same warranty as new parts. According to the data provided on the website of the group remanufactured in carbon-neutral factory the Kassel per year (VW 2012): 48,000 engines (in 490 versions), 60,000 cylinder heads (in 220 versions) and 49,000 transmissions (in 550 versions). In 2011, Volkswagen also began remanufacturing selected components in markets outside Germany and opened a new powertrain

Table 1 Design for remanufacture specifications (Amezquita et al. 1995)

D/W (D = Demand, W = Wish)		
1.	Materials	
W		All materials must be recyclable
W		No substantial increase in cost of materials
D		Must be corrosion resistant
D		Must be durable
D		Easily refurbishable
W		Light weight
W		Environmentally benign processing methods
D		Robust enough to reuse without replacement
W		Use recycled materials
D		Avoid toxic materials
D		Use secondary finishes such as painting, coating, etc.
W		Keep secondary finishes clear
2.	Assembly methods	
W		Less complex than existing methods
W		Faster than existing methods
W		Common method for diverse styles
D		Use Design for Assembly methods
W		Reduce number of components
3.	Fasteners	
D		Must be corrosion resistant
D		Must be durable
D		Must be reusable
D		Do not use screw heads which can be easily damaged (e.g., Torx, Phillips, etc.)
W		Do not combine metric and standard screws
D		Use standard fasteners
4.	Design for separability	
D		Choose joints that are easy to disassemble
D		Simplify and standardize component fits
W		Identify separation joints
W		Make adhesives safely soluble
W		Layout plastic parts close to top level of disassembly path
D		Provide "easy to see access" for disassembly
D		Provide access for power tool operation
5.	Cleaning	
D		Easy to handle and clean components
W		Do not use grooves or cavities that are hard to clean

(continued)

Table 1 Design for remanufacture specifications (Amezquita et al. 1995)

D/W (D = Demand, W = Wish)		
6.	Parts replacing	
D		Make parts susceptible to breakage easy to replace
W		Make parts susceptible to breakage separate from other parts
7.	Modular components	
D		Standard interfaces
D		Commonization/standardization of parts
8.	Design for recovery	
D		Parts must be high quality and durable
W		Parts must be easy to remove but not to steal

remanufacturing facility in China. This facility can remanufacture 15,000 engines a year to "as-new" quality.

Most of the automakers have remanufactured parts in their offer. It is a way to achieve win–win situation, when economics benefits can be combined with the fulfillment of legal regulations.

4 Recycling Strategy

Recycling is the process of collecting and processing materials destined to be thrown away as a rubbish into new products (EPA 2013). Recycling has a big influence at each stage of car life cycle, what is presented in Fig. 7.

At the designing stage there are made crucial decisions from recycling point of view. Designers and engineers are making restrictions of kinds of used materials, particularly there are chosen materials which are able to further processing—recyclable materials. At the same time using hazardous materials is avoided, because of difficulties for the manufacturer and the environment, too. Examples of design which takes into consideration recycling issues are presented in Table 2.

Utilizing recyclable materials causes possibilities of closed loop of used materials what minimize cost of manufacturing. For example Volkswagen has developed and introduced many processes ensuring at least 85 % recyclable and 95 % "recoverable" of his vehicles (VW 2011).

During manufacturing process there are given marks (according to requirements of ISO) what is useful at the next stages of parts' life. There are used recyclable parts chosen during the designing stage. Next stage—utilization is directly related to environment. There is a big problem—increasing number of old cars end of life, so there are some collecting points, which are collaborate with dismantling points

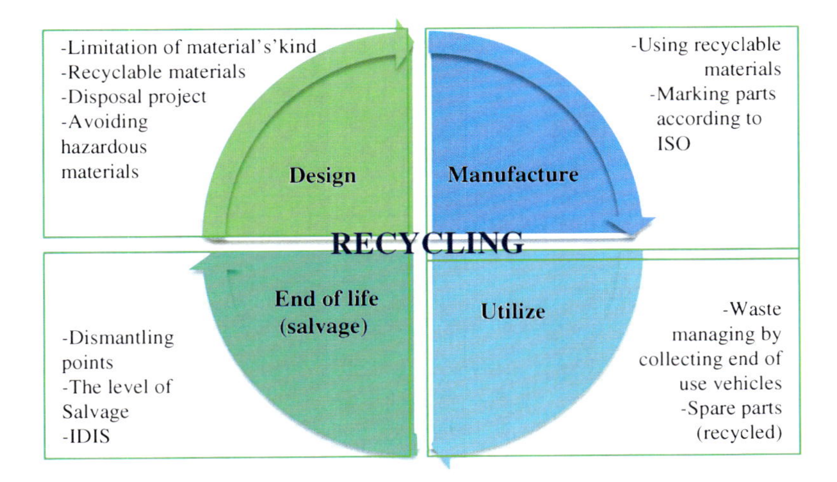

Fig. 7 Influence of recycling on car life cycle

Table 2 Designing influenced by recycling in Toyota company (Toyota 2011)

Category	Solution at designing affected by recycling
Innovation in materials	New Resin materials—Toyota Super Olefin Polymer (TSOP), a thermoplastic resin with better recoverability than the conventional reinforced composite polypropylene (PP) Example of use: the front and rear bumper of Corolla
Innovation in vehicle structure	Using the same type of thermoplastic resin for instrument panels, air-conditioning ducts, insulation pads, and sealing materials to improve the recoverability of dismantled parts. These parts are installed using friction welding rather than screws or metal clips, what eliminates the need for dismantling operations during recycling
Innovation in a sorting process	Material ID marking system (since 1981) to help identify materials used in resin parts. Currently, there is a marking system that conforms to international standards used for resin and rubber parts which weigh more than 100 g
Eliminating hazardous substances	Reduction of the amount of lead, which has been gaining attention as a substance of environmental concern in automobile shredder residue Example of parts without lead: battery cable terminals, copper radiators, copper heater cores, undercoating, high pressure hoses for power steering, side-protection moulding, wire harnesses, seat belt G sensors, fuel hoses

from the next stage. It becomes a huge problem to deal with the stream of old vehicles. To solve this problem there are created huge networks where end of live cars are collected and then dismantled to recycle.

At the stage called "End of life" there is a materials recovery. There are some level of salvage set by law regulation (for example in UE it is 95 % for salvage and

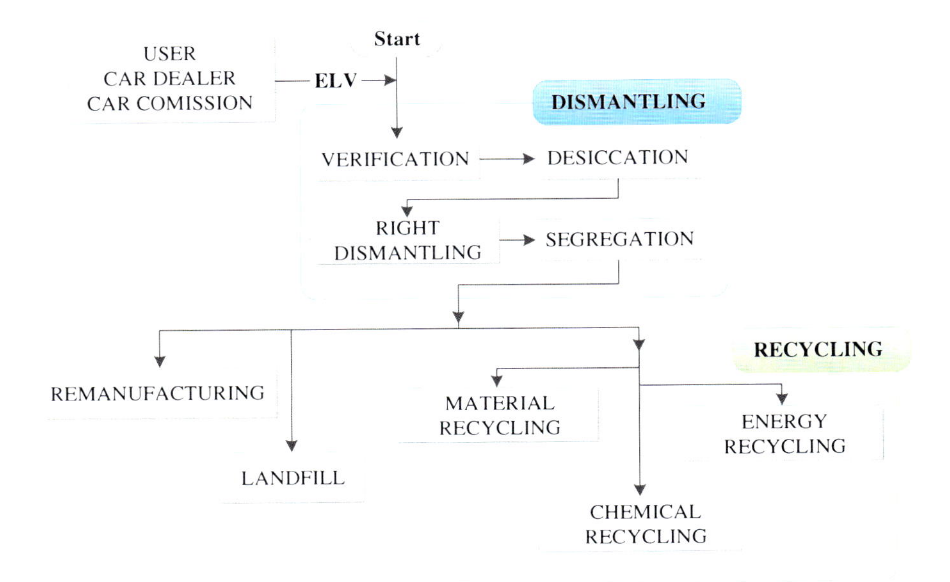

Fig. 8 The flow of ELV based on Lejda (2013), Lund (1998) and Nader and Jakowlewa (2009)

85 % for recycling according to ELV). Car manufacturers are striving to high level of salvage what is a result of law regulations,[1] but not only. From recycling point of view very important is IDIS (The International Dismantling Information System)—it was created by the automotive industry to meet the legal obligations of the ELV directive (IDIS 2013). The IDIS is a complex database with comprehensive information for each vehicle model relating to the removal of fluids such as fuel, oil and coolant, as well as the triggering of airbags and seat belt tensioners. It contains descriptions of the dismantling/removal of selected subassemblies and components, in particular for components with hazardous substances, such as batteries. It is free software for recovery and recycling companies (VW 2011).

Recycling process. Recycling process is very complex. ELV is a source of recycling process. In Fig. 8. authors present the flow of ELV taking into account the recycling process.

Car users, car dealers and car commissions are the source of the stream of ELV. All ELV regarding to law regulation, are allocated to dismantling process, where is made decision about the destiny of particular elements through the following activities:

(a) verification the condition of a part;
(b) drainage of all fluids

[1] In the UE—the main law regulation for automotive product end-of-life—Directive on End-of-Life Vehicle 2000/53/EC (Eur-lex 2013).

(c) right dismantling with the particular attention to dangerous substances such as brake pad, battery, which are removed;
(d) segregation into right groups which is a basis for the crucial decision about the destiny of an element.

After segregation components are divided into the groups which are remanufactured, disposed or recycled (Lejda 2013; Nader and Jakowlewa 2009).

There are the following possibilities of recycling (Lejda 2013):

1. Material recycling

The result of material recycling is obtaining materials able to using in a manufacturing process. The process of material recycling includes melting scrap metal, pulverization of plastic material elements etc. The most important issues in material recycling are:

- recovery of the same compounds,
- manufacturing of new compounds,
- re-construction of new car parts,
- manufacturing of other commercial products.

The simplest example is metal recycling process—for example steel, which in the recycling process cannot achieve the same quality parameters as before, but it can be used for manufacturing of less important car parts or elements used in building construction. Thermoplastic materials, e.g. polypropylene, polyethylene, polyurethane materials etc. are well-adapted for recycling.

For example—the Mercedes company recovers about 90 % of used plastic materials and they produce multi-layered bumpers, sound-dumping screens, air ducts, electric cross-cable tubes, hub cover caps, chassis protection shields etc. Another example is Renault company, which has set an objective of using 50 kg recycled plastic per car in 2015, equal to an average 20 % of all plastic content.

2. Energy recycling

Energy recycling includes not only burning waste, but also generating fuels from them and processing them to create the thermal insulation materials. Examples of energy recycling includes: producing energy for water heating, steam and electricity generation, etc. Tires are the most important item in energetic recycling of car.

3. Chemical recycling

Chemical recycling consists of thermal and chemical processing of elements made of plastics into elementary compounds such as gases or oils which are used as chemical components or as fuels. Chemical recycling is also processing of used-up motor oils and other dispensable fluids, for example brake fluid is processed into household cleaning chemicals.

Car which is ELV is decomposed into prime factors. There can be distinguished five main material fraction (Oleszczuk 2003):

Table 3 Technologies of material recycling elaborated by Toyota company (Toyota 2011)

Type of material	Original form	Recycled part
Thermoplastic resin	• TSOP (Toyota Super B Olefin Polimer) bumper	• Bumper, fuel tank protector Luggage trim, Fuel pump protector, Seat backboard, Seat under-cover, Lamp cover, Back door trim cover, Engine under-cover, Luggage compartment trim, Bumper step, Deckside trim
	• Interior trim, garnish	• Timing belt cover • Fan shroud
Thermosetting resin	• Parts from reinforced plastic FRP	• Sunroof housing • Cylinder head cover
Resin composite material	• Carpets • Seat covers • Instrument panel covering • Molded roof lining	• Carpet backing • Carpet reinforcement parts • Floor silencer • Dash silencer • Interior trank
Rubber	• Weather stripping • Tire	• Hose protector • Weather stripping • Alternative fuel for cement
Automobile shredder Residue (ASR)	• Urethane foam and fiber • Copper wiring • Glass	• Damping materials • Reinforcing materials for aluminium casting • Reinforcing materials for tiles
Steel	• Body, door	• Car parts, general steel product
Other materials	• Bottles from PET	• Materials suppresses noise

- rubber and plastic
- ferrous
- nonferrous metal
- glass
- plastic.

Car consists of: different kind of plastic, glass, ferrous and nonferrous metal and rubber/tires. Each of material can be recycled. Possibilities of recycling are the result of: available technology of processing, demand on remanufactured parts and economic relationship between new part and recycled one.

Toyota company is very advanced car manufacturer at the recycling. It is a leader in the use of recycled materials for several years. In Table 3 are presented some of examples of technologies of material recycling used in Toyota company.

Recycling is not a possibility of energy, cost and material savings but it is a necessity to deal with limited resources. Car manufacturers try to build up recycling strategy also in relations with their suppliers practice. In last decade a number of sustainable practice was introduced like e.g. issuing recycling guidelines to the suppliers.

5 Conclusion

The number of cars is growing every year, as well as numbers of cars which are not any more in use and need recycling. Automotive industry became a big load for the environment, but there is no chance to stop it. We can only decrease negative influence, what car makers are already doing. There are 3 strategies in automotive industry related to environment: reducing, reusing and recycling strategy, which are differed from each other, but all of them are taking into account the environment protection.

All strategies should be considered by the point of view of different sides of automotive industry including: car makers (processes and employees at each stage of car life cycle), car users, suppliers, etc.

There are presented many possibilities of realizing 3R strategy in the automotive industry but authors indicate that protection of the environment and all activities related to sustainable development is not a presentation of higher awareness of people. The source of interest in environmental issues is law regulations and profitability—possibility of reducing cost by car manufacturers. The environmental issues will be more popular not only due to economic benefits but also due to higher environmental awareness among car manufacturer, suppliers, car users and all participants of so called "automotive world".

References

Amezquita T, Hammond R, Bras B (1998) Issues in the automotive parts remanufacturing industry—a discussion of results from surveys performed among remanufacturers. Int J Eng Design Autom Special Iss Environ Conscious Design Manuf 4(1):27–46

Amezquita T, Hammond R, Marc M, Bras B (1995) Characterizing the remanufacturability of engineering systems. Massachusetts, DE-Vol 82, pp 271–278

APRA (2013) http://www.apra-europe.org/main.php?target=availableproducts

Directive 2005/64/EC of the European Parliament and of the Council of 26 October 2005 on the type-approval of motor vehicles with regard to their re-usability, recyclability and recoverability and amending Council Directive 70/156/EEC

EPA (2013) http://www.epa.gov/recycle/recycle.html

Eur-lex (2013) http://eur-lex.europa.eu/LexUriServ/LexUriServ.do?uri=CELEX:32000L0053:EN:NOT

FIAT (2011) Sustainability report. http://www.fiatspa.com/en-US/sustainability/enviromental_responsibility/logistics/Pages/environmental_performance.aspx

Guide VDR Jr (2000) Production planning and control for remanufacturing: industry practice and research needs. J Oper Manag 18(4):467–483

Henstock ME (1988) Design for recyclability. Institute of Metals, London

IDIS (2013) http://www.idis2.com/index.php?&language=english

Lejda K (2013) Selected problems in car recycling. http://www.pan-ol.lublin.pl/wydawnictwa/TMot4/Lejda.pdf

Lund R (1983) Remanufacturing: United States experience for developing nations. The World Bank, Washington, DC

Lund R (1998) Remanufacturing: an American resource. In: Proceedings of the 5th international congress on environmentally conscious design and manufacturing, June 16–17, Rochester Institute of Technology, Rochester, NY

Nader M, Jakowlewa I (2009) Selected aspects of the organization of car recycling plant. Scientific papers of the University of Warsaw, z.70/2009, Warsaw, pp 127–138

Oleszczuk P (2003) Characteristics and possible ways of disposing of the residue after crushing a car. Protection Air Waste Probl 5(2003):151–156

Pahl G, Beitz W (1984) Engineering design (trans: Wallace K). The Design Council/Springer, London/Berlin

Parkinson HJ (2001) Systematic approach to the planning and execution of product remanufacture. PhD thesis, Department of Mechanical Engineering, UMIST

Parkinson HJ, Thompson G (2003) Analysis and taxonomy of remanufacturing industry practise. Proc Inst Mech Eng E J Process Mech Eng 217(3):243–256

REMAN (2013) http://www.reman.org/pdf/steinhilper_part4.pdf

Steinhilper R (1998) Remanufacturing: the ultimate form of recycling. Fraunhofer IRB. Verlag, Stuttgart

Severengiz S, Sezgin C, Yonk NK (2008) Automotive remanufacturing in Europe: survey results. TU Berlin

Toyota (2011) Car recycling brochure. http://www.toyota.de/Images/Car_Recycling_Brochure_EN_tcm281-210236.pdf

VW (2011) Sustainability report. http://sustainability-rport2011.volkswagenag.com/fileadmin/download/71_Recycling_Sicon.pdf

VW (2012) http://nachhaltigkeitsbericht2012.volkswagenag.com/en/environment/resource-efficiency/vehicle-recycling.html

A Declarative Approach to New Product Development in the Automotive Industry

Marcin Relich

Abstract The chapter aims to present a declarative approach for management of the new product development project portfolio in the automotive industry. A reference model of an automotive company and project portfolio planning is formulated in terms of a constraint satisfaction problem and implemented in constraint programming languages, facilitating the development of a decision support system that seeks a feasible set of alternatives for project portfolio completion. It is especially attractive in the case of a lack of possibility for continuing the baseline schedule, and supports managers in choosing an alternative schedule. As a consequence, project portfolio management is more efficient, the competitiveness of the automotive company increases, and the launch of new vehicle models containing technologies less harmful to the environment is faster. The chapter includes illustrative examples concerning new sustainable trends in the automotive industry.

Keywords Reference model · Constraint satisfaction problem · Project portfolio planning · Decision support system · Sustainability

1 Introduction

The automotive industry is one of the most important worldwide drivers of growth and employment, as well as technological and managerial innovation. Due to an increasingly complex and competitive global marketplace, automotive companies are looking at new ways to improve their operation in order to remain profitable. In particular, premium class production companies have to launch their innovations

M. Relich (✉)
Faculty of Economics and Management, University of Zielona Góra,
Licealna 9, 65-417 Zielona Góra, Poland
e-mail: m.relich@wez.uz.zgora.pl

P. Golinska (ed.), *Environmental Issues in Automotive Industry*,
EcoProduction, DOI: 10.1007/978-3-642-23837-6_2,

quickly and to achieve a short time to market in order to gain competitive advantages and to underline their premium image (Grimm 2003). One of the characteristics of the automotive industry concerns the management of several simultaneously developed new products (projects). Automotive companies usually have several product lines and constantly develop new products to replace existing products or to add completely new product lines.

The development of a new car is based on long running, concurrently executed and highly dependent product lines. Different life cycle times of mechanical, software and hardware components as well as differing product development durations requires efficient coordination. Furthermore, product-driven process structures, dynamic adaptation of these structures, and handling real-world exceptions result in challenging demands for any IT system (Müller et al. 2006). The planning, monitoring, and evaluation of portfolios in an automotive company can be supported by the functionalities of an integrated information system (e.g. Enterprise Resource Planning (ERP)) or a stand-alone application (e.g. Microsoft Project). The advantage of the use of an integrated system in the field of project management lies in the use of a single database with online access to resource levels, company calendars, etc. In turn, by using a stand-alone application for project management, the ERP system is not so overloaded and it can manage the routine processes in an enterprise such as inventory, order entry, shipping, supply more efficiently.

Present ERP systems tend to reflect the vast field of business processes in an enterprise, as well as storing data concerning, for example, sales and purchasing transactions. However, some processes concerning unstructured problems are very difficult in implementation and rarely occur in the ERP system. For instance, in the automotive company an unstructured problem can concern product development, given the enormous number of scenarios for project portfolio completion. One of the reasons behind the difficulty in such situations is that current business process modelling languages and models are of imperative nature—they prescribe strictly how to work. The imperative nature of modelling languages forces the designer to over-specify processes, resulting in frequent changes (Pesic and Aalst 2006). Moreover, the imperative nature of modelling leads to the classical drawbacks of being forced to discard degrees of freedom and side constraints. Discarding degrees of freedom may result in the elimination of interesting solutions, regardless of the solution method used. Discarding side constraints gives a simplified problem and solving this simplified problem may result in impractical solutions for the original problem (Rossi et al. 2006).

The limitations of imperative languages provide the motivation to develop a reference model of project management in an automotive company and to implement it in declarative languages. Unlike imperative languages, declarative languages specify the "what" without determining the "how". The advantage of working with such a model is that users are driven by the system to produce the required results, whilst the manner in which the results are produced is dependent on the preferences of the users (Pesic and Aalst 2006).

The project portfolio scheduling problems belong to the class of NP-complete problems. Several methods and techniques have been proposed in this field, starting from computer simulation that has been used as a model component in building model-driven decision support systems (Power and Sharda 2007), through operational research including MILP (Sawik 2007), dynamic programming (Lewis and Slotnick 2002), branch-and-bound (Burke and Kendall 2005) to artificial intelligence solutions (Bocewicz et al. 2009). The inherent uncertainty and imprecision in project scheduling have motivated the proposal of several fuzzy set theory based extensions of activity network scheduling techniques (Maravas and Pantouvakis 2012). Among these extensions, for instance, fuzzy variables in PERT (Chanas and Kamburowski 1981), resource-constrained fuzzy project-scheduling problem (Bonnal et al. 2004), criticality analysis of activity networks with uncertainty in task duration (Fortin et al. 2010), interval PERT and its fuzzy extension (Dubois et al. 2010) and fuzzy repetitive scheduling method (Maravas and Pantouvakis 2012) can be found. The extension of the proposed approach towards the presence of uncertainty is subject to further research.

In the field of constraint-based scheduling two strengths emerge: natural and flexible modelling of scheduling problems as Constraints Satisfaction Problems (CSP) and powerful propagation of temporal and resource constraints. Thus, the scheduling problem is modelled as CSP at hand in the required real-life detail. The proposed approach aims at elaborating a reference model of project portfolio in an automotive company, specifying this model in terms of CSP, and using constraint programming (CP) to seek a solution to the problem. The scheduling and car sequencing problem is one of the classical problems presented in CSP literature (e.g. Dincbas et al. 1988; Brailsford et al. 1999; Badra et al. 2011). However, the application of CP in new automotive product development has not yet been considered.

The presented approach concerns the phase of product design and development that includes component and integration tests, and therefore is crucial for further phases of vehicle life cycle. With global vehicle usage expected to increase in the next decades, the impact on global air quality, human health, and global climate could be extremely damaging if significant changes in vehicle design are not implemented globally to reduce these negative trends (Mayyas et al. 2012). The shift towards more sustainable constructions in the automotive industry is not only an initiative towards a more viable environment and cost efficiency but also the demand of European regulations (Koronis et al. 2013). These regulations are an important factor for the development of sustainable technologies that reduce CO_2 emission and energy consumption, including the use of alternative fuels, new engine technologies, reducing rolling resistance, improving vehicle aerodynamics, and reducing vehicle weight (Mayyas et al. 2012). Some of selected up-to-date technologies, that may offer very promising potential to provide benefits to companies, the natural environment and end-customers are considered in an illustrative example in Sect. 5.

In this context, the contribution covers various issues of decision-making while employing the knowledge and CP based framework. The model formulated in terms of CSP (as a descriptive approach) determines a single knowledge base and it enables effective implementation in constraint programming languages, as well as the development of a task-oriented decision support system for automotive project portfolio planning. As a result, the problem specification is closer to the original problem, obtaining solutions that are unavailable with imperative programming. Consequently, project portfolio management is more efficient and automotive companies can launch new vehicle models containing technologies less harmful to the environment much more quickly.

The remaining sections of this chapter are organised as follows: Sect. 2 presents a reference model that encompasses the characteristics of an automotive company and project management. A problem formulation in terms of CSP for automotive project portfolio planning is specified in Sect. 3. A method for obtaining alternative variants of project portfolio is shown in Sect. 4. An illustrative example of the approach, which presents the possibility of decision problem specification in a direct and in an inverse way and considers up-to-date environmentally friendly technologies, is presented in Sect. 5. Finally, some concluding remarks are contained in Sect. 6.

2 A Reference Model of an Automotive Company and Project Management

The results of the research indicate very strongly that the version of project management depends on the characteristics of the company. For instance, small to medium enterprises require less bureaucratic forms of project management than those used by larger, traditional organizations (Turner et al. 2010). This provides motivation to consider project management in connection with the nature of an organization and to develop a reference model that is tailored to the automotive company.

The emergence of the systems approach to management has had a significant influence on organizational theory and management philosophies. The systems approach is a dynamic process that integrates all activities into a meaningful total system, and it seeks an optimal solution or strategy in solving a problem (Kerzner 2009). An organization considered as a system consists of the subsystems such as sales, marketing, supply, manufacturing, inventory, accounting, human resources (HR), information, R&D, etc. The activity of an organization also depends on its external environment such as the nature, economical, political, technological, and social conditions, as well as competitors, clients, suppliers (also sub-contractors), and availability of resources (including human). An illustration of an organization and its environment in terms of the system theory is presented in Fig. 1.

In most enterprises, functions concerning resource control, decision-making, sales and cost planning, employee motivation, etc. occur. These functions can be considered as a chain of processes, the execution of which is connected with the

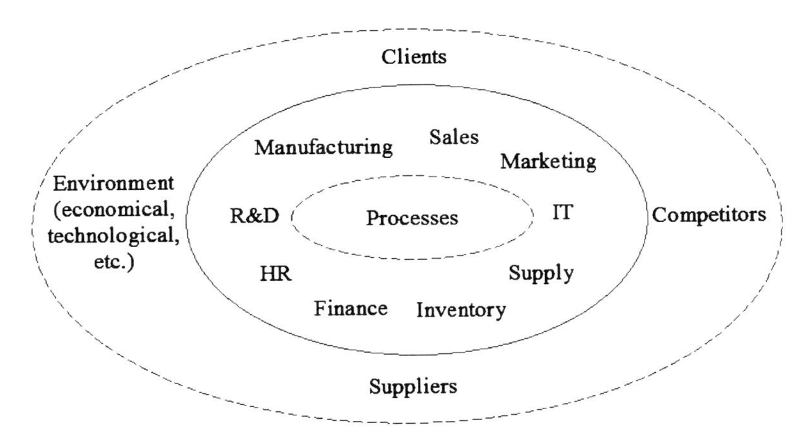

Fig. 1 Hierarchical model of an organization

organization's characteristics and its environment. For instance, cost planning requires information about the number of client orders, price of available resources, software for planning support (with different techniques of parameter estimation), communication between employees, etc. In the case of product development in an automotive company, the information is obtained from different subsystems of the organization. An example of information supporting decision-making in the project portfolio planning process, in connection with the organization's subsystems, is presented in Table 1.

Project management is classified in various ways in literature. According to the Project Management Institute, project management consists of nine knowledge fields (PMI 2008). Another classification comprises 17 elements of project management (Turner 2006). Despite the diverse classification of project management, it usually concerns elements such as time, cost, resources, risk, and communications (see Fig. 2). Each of the fields includes the subsequent elements that can be considered as the processes of project management. For instance, cost management consists of estimating, budgeting, and controlling costs in a project (PMI 2008).

Table 1 Information in project portfolio planning process

Organization's subsystem	Information
Supply	Cost of materials for a prototype
Manufacturing	Time and cost of prototype production
Sales	Demand for the previous versions of cars
Marketing	Market research concerning a developed version of a car
Inventory	Availability of components for a prototype
Finance	Cost accounting in the different dimensions
HR	Possibility of new employee recruitment
R&D	Past experiences in development of new car versions

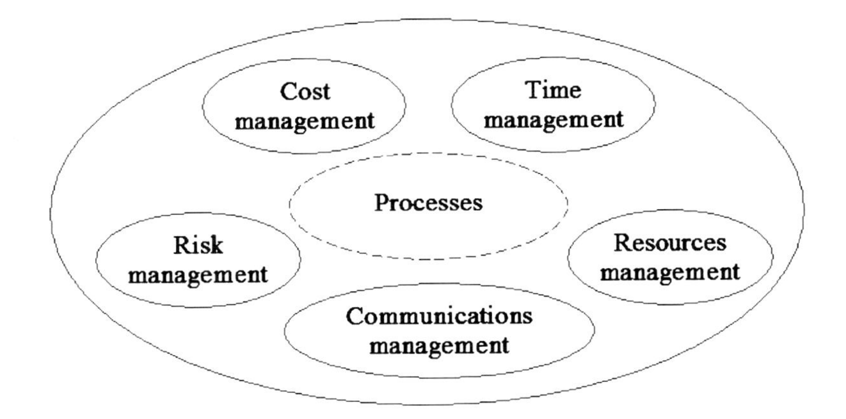

Fig. 2 Hierarchical model of project management

The processes in both the above-mentioned fields can be described by a set of variables and constraints, and combined by the relationships occurring between them (constraints—C), e.g. the financial means in a company must be greater than project budget. Figure 3 depicts an example of a proposed reference model for an organization and project management.

The processes of organization management contain variables (e.g. employees, equipments, vehicles), and the constraints concern the level of resources, their availability in time, etc. Similarly, it is also possible to consider the relationships between different processes in project portfolio management. The variables of project portfolio include e.g. set of activities, their duration, and starting time. In turn, the relations include precedence constraints, horizon of project portfolio, positive net present value, etc.

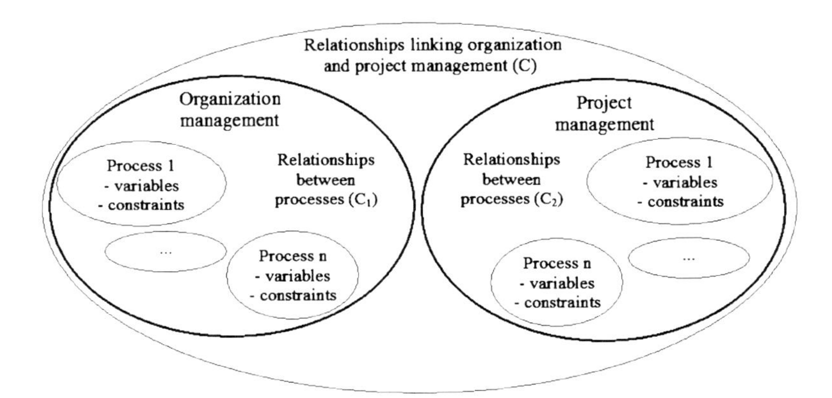

Fig. 3 Framework of proposed reference model

The constraints combine and limit the decision variables describing the capacity of an enterprise as well as these connected with project execution. For instance, the number of enterprise employees limits the time of project implementation. Consequently, the fulfilment of assumed constraints enables project execution within the organization according to the requirements. This approach seems to be natural in the case of an organization that executes projects and solves standard decision-making problems. In this case, a knowledge base is created that, in addition to the inference strategies, allows more efficient implementation of a decision support system. In this context, the problem statement of project planning, specified in terms of a constraint satisfaction problem, is presented in the next section.

3 Problem Statement of Project Portfolio Planning in CSP

The presented model contains a set of decision variables, their domains, and the constraints. Hence, it seems natural to describe a planning problem in terms of a constraint satisfaction problem (*CSP*). The model description encompasses the assumptions of an enterprise, implementing projects therein, and a set of routine queries (the instances of decision problems) that are formulated in the framework of *CSP*. The structure of the constraint satisfaction problem may be described as follows (Rossi et al. 2006):

$$CSP = ((V, D), C) \tag{1}$$

where:

$V = \{v_1, v_2, \ldots, v_n\}$—a finite set of n variables,
$D = \{d_1, d_2, \ldots, d_n\}$—a finite set of n discrete domains of variables,
$C = \{c_1, c_2, \ldots, c_m\}$—a finite set of m constraints limiting and linking variables.

The hierarchical structure of the reference model implies a similar structure concerning the constraint satisfaction problem (Fig. 4). The reference model can be described as a single *CSP* that consists of CSP_1 and CSP_2 concerning the fields of enterprise and project management, respectively. In turn, CSP_1 and CSP_2 contain successive elements describing the processes of different fields ($CSP_{11}, \ldots, CSP_{1n}$ and $CSP_{21}, \ldots, CSP_{2n}$).

According to the formula (1) CSP_1 can be described as follows:

$$CSP_1 = ((R, D_R), C_{CSP1}) \tag{2}$$

where:

$R = (r_{1,1}, \ldots, r_{k,h}, \ldots, r_{K,H})$—the k-th discrete resources (e.g. workforce, tools, financial means) in the h-th time unit ($h = 0, 1, \ldots, H$);

D_R—a set of an admissible amount of R resources;

C_{CSP1}—a set of constraints (e.g. a number of available employees and financial means in the enterprise).

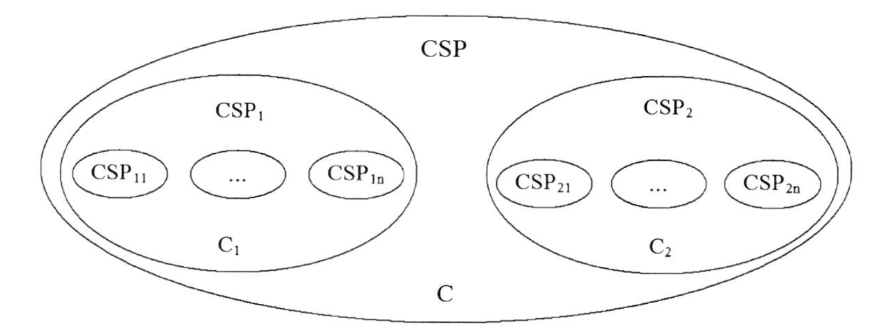

Fig. 4 Structure of constraint satisfaction problem

Among the project management processes (CSP_2), a scheduling problem is further considered. The extension of the proposed model concerning other fields of project management is subject to further research. Given a set of projects $P = \{P_1, P_2, \ldots, P_I\}$, where the project P_i consists of J activities: $P_i = \{A_{i,1}, \ldots, A_{i,j}, \ldots, A_{i,J}\}$. It is assumed that an activity is indivisible and an activity can start its execution only if the required amount of resources are available at the moments given by $Tp_{i,j}$ and after completion of the previous activity. The resource can be allotted or released only after completion of the activity that requires this resource. The project P_i is described as an activity-on-node network, where nodes represent the activities and the arcs determine the precedence constraints between activities.

The CSP_2 can be described as follows:

$$CSP_2 = ((\{P_i, A_{i,j}, s_{i,j}, t_{i,j}, Tp_{i,j}, Tz_{i,j}, Dp_{i,j}\}, \{D_{Pi}, D_{Ai}, D_{si}, D_{ti}, D_{Tpi}, D_{Tzi}, D_{Dpi}\}), C_{CSP2}) \tag{3}$$

where:

P_i—i-th project;

$A_{i,j}$—j-th activity of i-th project that is specified following: $A_{i,j} = \{s_{i,j}, t_{i,j}, Tp_{i,j}, Tz_{i,j}, Dp_{i,j}\}$;

$s_{i,j}$—the starting time of the activity $A_{i,j}$, i.e., the time counted from the beginning of the time horizon H;

$t_{i,j}$—the duration of the activity $A_{i,j}$;

$Tp_{i,j} = (tp_{i,j,1}, tp_{i,j,2}, \ldots, tp_{I,J,K})$—the sequence of allocation moments by the activity $A_{i,j}$ required resources: $tp_{i,j,k}$—the time counted since the moment $s_{i,j}$ of a number $dp_{i,j,k}$ of the k-th resource allocation to the activity $A_{i,j}$. That means the k-th resource is allotted to an activity during its execution period: $0 \leq tp_{i,j,k} < t_{i,j}$; $k = 1, 2, \ldots, K$;

$Tz_{i,j} = (tz_{i,j,1}, tz_{i,j,2}, \ldots, tz_{I,J,K})$—the sequence of moments, when the activity $A_{i,j}$ releases the resources: $tz_{i,j,k}$—the time counted since the moment $s_{i,j}$ of a number $dp_{i,j,k}$ of the k-th resource release by the activity $A_{i,j}$. That means the k-th resource is released by activity during its execution period: $0 < tz_{i,j,k} \leq t_{i,j}$; $tp_{i,j,k} < tz_{i,j,k}$;

$Dp_{i,j} = (dp_{i,j,1}, dp_{i,j,2}, \ldots, dp_{I,J,K})$—the sequence of number of the k-th resource is allocated to the activity $A_{i,j}$: $dp_{i,j,k}$—a number of the k-th resource allocation to the activity $A_{i,j}$;

D_{Pi}—a set of admissible number of projects in an enterprise;

D_{Ai}—a set of admissible number of activities in the i-th project;

D_{si}—a set of admissible starting times of activity $A_{i,j}$ in the i-th project;

D_{ti}—a set of admissible durations of activity $A_{i,j}$ in the i-th project;

D_{Tpi}—a set of admissible allocation moments to activity $A_{i,j}$ for the k-th resource in amount of $dp_{i,j,k}$, in the i-th project;

D_{Tzi}—a set of admissible release moments by activity $A_{i,j}$ for the k-th resource in amount of $dp_{i,j,k}$, in the i-th project;

D_{Dpi}—a set of admissible number of required resources by the activity $A_{i,j}$ in the i-th project;

C_{CSP2}—a set of constraints:

$C_{CSP2,1}$—constraint concerning horizon of project completion:

$$\forall s_{i,j}\, \forall t_{i,j}\, (s_{i,j} + t_{i,j} \leq H) \tag{4}$$

$C_{CSP2,2}$—precedence constraints:

- the n-th activity follows the i-th one:

$$s_{i,j} + t_{i,j} \leq {}_{i,n} \tag{5}$$

- the n-th activity follows other activities:

$$\begin{aligned} s_{i,j} + t_{i,j} &\leq s_{i,n} \\ s_{i,j+1} + t_{i,j+1} &\leq s_{i,n} \\ &\ldots \\ s_{i,j+n} + t_{i,j+m} &\leq s_{i,n} \end{aligned} \tag{6}$$

- the n-th activity is followed by other activities:

$$\begin{aligned} s_{i,n} + t_{i,n} &\leq s_{i,j} \\ s_{i,n} + t_{i,n} &\leq s_{i,j} + 1 \\ &\ldots \\ s_{i,n} + t_{i,n} &\leq s_{i,j} + m \end{aligned} \tag{7}$$

In the reference model of project portfolio planning, the constraints C are the elements linking CSP_1 and CSP_2 (see Fig. 4) and they can be as follows:

C_1—the total number of employees $r_{1,h}$ in the enterprise in the h-th time unit cannot be less than a number of employees $r_{1,h,i}$ for project portfolio:

$$r_{1,h,i} \leq r_{1,h} \tag{8}$$

C_2—the total financial means $r_{2,h}$ in the enterprise in the h-th time unit cannot be less than the financial means for project portfolio $r_{2,h,i}$:

$$r_{2,h,i} \leq r_{2,h} \tag{9}$$

CSP is implemented according to the structure of the reference model, and can be also considered as a knowledge base. The knowledge base is a platform for query formulation as well as for obtaining answers, and it comprises of facts and rules that are characteristic of the system's properties and the relations between its different parts. As a consequence, a single knowledge base facilitates the implementation of a decision support system.

A knowledge base can be considered in terms of a system. At the input of the system are the variables concerning basic characteristics of an object that are known and given by the user. For instance, the variables concerning available resources in the enterprise and a sequence of project activities occur in the knowledge base describing the enterprise-project model. The output of the system is described by the characteristics of the object that are unknown or are only partially known. In the considered case, the variables include the cost and time of an activity as well as the resources usage.

A distinction of decision variables that are embedded in the knowledge base as an input–output variable permits the formulation of two classes of standard routine queries that concern two different problems with respect to resources (Jüngen and Kowalczyk 1995):

- a fixed (limited) amount of resources is available: what are the consequences of these resource constraints for the execution of the project?
- what resources in which quantities are minimally necessary to be able to finish the project before a certain deadline?

The first type of above routine query is further called the direct approach and the second the inverse approach. These categories of questions encompass the different reasoning perspectives, i.e. deductive and abductive approaches. The method concerning the determination of admissible solutions for the above-described problem is presented in the next section.

4 Method for Obtaining Feasible Variants of Project Portfolio

Every method is dedicated to (i.e. the most effective for) a specific class of problems, and its effectiveness can be understood as a manner dictated by the nature of query and the specifics of the assumed model (Banaszak and Bocewicz 2011). In the case of the considered problem, the method consists of the following stages:

1. seeking a problem solution (i.e. project portfolio schedule) for given variables, their domains, and constraints (a direct approach); if there is no solution (e.g.

the planned project completion exceeds the target time), then the next stage is conducted;
2. seeking the values of decision variables (e.g. connected with resource reallocation in the enterprise) that satisfy the assumed project constraints (an inverse approach);
3. if there is a set of solutions, then the optimal variant of alternative project portfolio completion is chosen according to a criterion for assessment, e.g. time and/or cost minimisation.

The planning issue and subsequent monitoring of the project is one of the most important elements of project management that determines its success or failure (Kerzner 2009). If the planning indicates an overrun of constraint (e.g. target cost), and there are no possibilities to increase the constraint (e.g. by obtaining the additional financial means, contract renegotiation), then such the values of variables for which the assumed constraints are satisfied are sought. A choice of variable(s) for adjustment depends on the problem considered and availability of resources.

In the proposed approach, monitoring costs and activity performance, as well as re-planning project completion (with regards to schedule, resource usage) follows in time unit h (e.g. weekly, monthly interval). The actual cost concerning partial project performance (solid line) as well as an example of an alternative variant of project completion (dashed line) is presented in Fig. 5.

If the planned total cost is greater than the target cost (basic variant), then an alternative variant is sought that fulfils the assumed constraints. The alternative variant is one of the feasible solutions, which is chosen according to a criterion for assessment. Often, there is a trade-off between the duration of an activity and the cost connected with the execution of the activity. To each activity, a cost function can be assigned which represents the cost of the activity given certain duration of that activity. The criterion can be for example one of the following (Jüngen and Kowalczyk 1995):

- to minimize the total cost of the project portfolio,
- to minimize the total duration of the project portfolio,
- to find the minimal total cost of the project portfolio, given its maximum acceptable total duration.

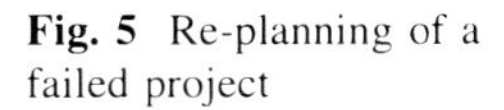

Fig. 5 Re-planning of a failed project

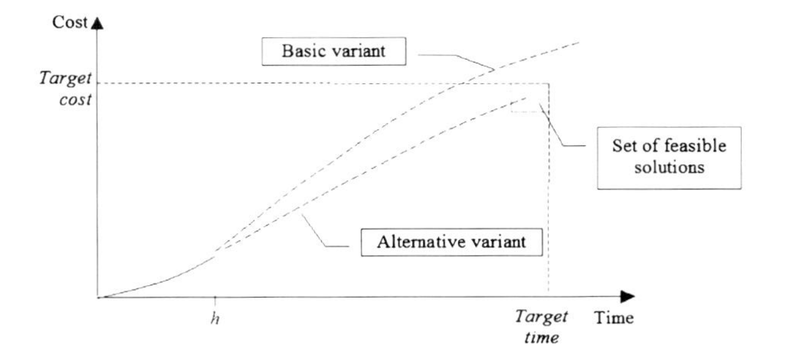

The proposed model enables a descriptive approach to the problem statement, which encompasses constraint satisfaction problem structure and then allows the implementation of the considered problem in the constraint programming environment. CP is an emergent software technology for declarative CSP description and can be considered as a pertinent framework for the development of decision support system software aims. A CSP can always be solved with a brute force search. All possible values of all variables are enumerated and each is checked to see whether it is a solution. However, for many intractable problems, the number of candidates is usually too large to enumerate them all. CP has developed some ways (constraint propagation and variable distribution) to solve CSP that greatly reduce the amount of search needed. Constraint propagation can notably simplify a constraint problem and so improve the efficiency of a search for a solution. In turn, backtracking search algorithms come with a guarantee that a solution will be found if one exists, and can be used to show that a CSP does not have a solution and to find a provably optimal solution (Rossi et al. 2006). These algorithms are sufficient for solving many practical problems (e.g. project portfolio planning and scheduling). CP is qualitatively different from the other programming paradigms, in terms of declarative, object-oriented and concurrent programming. Compared to these paradigms, constraint programming is much closer to the ideal of declarative programming: to say what we want without saying how to achieve it (Van Roy and Haridi 2004; Relich 2011).

The seeking of feasible solutions for an inverse approach is connected with verification of an available change for values of variables, constraints, and/or assessment criterion. For instance, they can include obtaining new resources for the execution of critical project activities and resource reallocation between a number of ongoing projects. An example concerning the considered problem described in the constraint programming environment is presented in the next section.

5 Illustrative Example

The example consists of a project portfolio description and the formulation of a project portfolio planning problem in the direct and inverse approach. The presented project portfolio contains examples concerning up-to-date environmentally friendly technologies such as LED lighting, a new electric motor, and lightweight and sustainable green materials.

5.1 Project Portfolio Description

The development of new products in the automotive industry (parts for a new version of a car) is usually simultaneously conducted and may be considered as a project portfolio. The structure of presented projects takes into account the automotive quality management system international standard (ISO/TS16949), which is related to the product realization process and includes the following phases: preliminary research stage, product development, production launch, feedback assessment and corrective action.

The first project (P_1) concerns a redesign of a car body and the development of new LED technology for the lamps. The new design of a car is usually a steady policy of an automotive company, which results from customer requirements and the force of competitors and suppliers. The application of LED technology significantly contributes to the avoidance of CO_2 emission as well as the reduction of fuel and energy consumption, which is advantageous from both the customer and the environmental issues point of view. Moreover, LED headlights do not contain hazardous waste and are recyclable, eliminating pollution and the danger of related mercury absorption into the environment. Recently, various technological improvements have been occurred that have made a major contribution to increasing LED lighting uptake in the future, as they reduce the number of LED chips needed, reduce overall lighting costs, reduce power consumption, and improve LED usage rates. The description of activities in project P_1 is presented in Table 2.

The network diagram of the activities in the project P_1 is shown in Fig. 6. The project completion time equals 23 months.

The second project (P_2) is connected with the development of a new electric motor, a battery, and other required electric components. A new class of high-efficiency electric motors enables highly efficient, more cost effective and less

Table 2 Description of activities in project P_1

Activity	Description	Duration (months)	Predecessor(s)
$A_{1,1}$	Research plan development	2	–
$A_{1,2}$	Assessment of legal regulations, patents, and environmental issues	1	$A_{1,1}$
$A_{1,3}$	Initial financial and risk analysis	1	$A_{1,2}$
$A_{1,4}$	R&D in longer lifespan and heat dissipation of LED products and design of lamps according to LED technology	6	$A_{1,3}$
$A_{1,5}$	R&D in raising the luminous efficacy of LEDs	2	$A_{1,3}$
$A_{1,6}$	R&D in easier use of LED products with mainstream electric grids	2	$A_{1,3}$
$A_{1,7}$	Design of car body and cockpit	6	$A_{1,3}$
$A_{1,8}$	Drafting specifications and supplier sourcing	6	$A_{1,4}$, $A_{1,5}$, $A_{1,6}$, $A_{1,7}$
$A_{1,9}$	Trial preparation	1	$A_{1,4}$, $A_{1,5}$, $A_{1,6}$, $A_{1,7}$
$A_{1,10}$	Integration tests and validation	4	$A_{1,8}$, $A_{1,9}$
$A_{1,11}$	Production test series	3	$A_{1,10}$

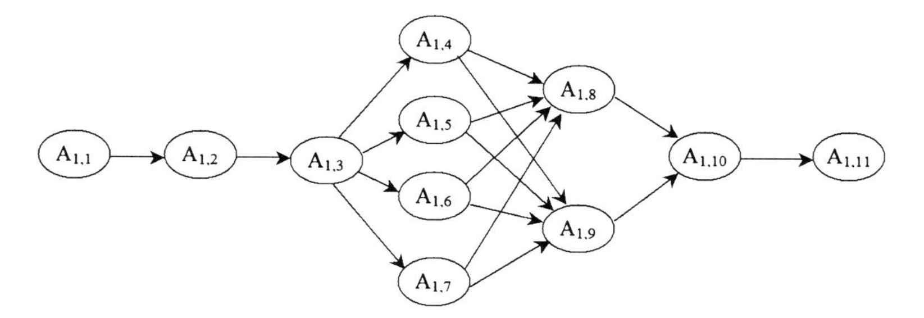

Fig. 6 Network diagram for project P_1

Table 3 Description of activities in project P_2

Activity	Description	Duration (months)	Predecessor(s)
$A_{2,1}$	Project planning	2	–
$A_{2,2}$	Concept development	2	$A_{2,1}$
$A_{2,3}$	Assessment of legal regulations, environmental issues, and risk analysis	1	$A_{2,2}$
$A_{2,4}$	Increment of electric motor acceleration	9	$A_{2,3}$
$A_{2,5}$	Battery technology improvement	6	$A_{2,3}$
$A_{2,6}$	Improvement of controller performance	3	$A_{2,3}$, $A_{2,4}$, $A_{2,5}$
$A_{2,7}$	Virtual prototype	6	$A_{2,4}$, $A_{2,5}$, $A_{2,6}$
$A_{2,8}$	Analysis and selection of suppliers	2	$A_{2,4}$, $A_{2,5}$, $A_{2,6}$
$A_{2,9}$	Trial version of the products	6	$A_{2,7}$, $A_{2,8}$
$A_{2,10}$	Analysis of the trials	4	$A_{2,9}$

resource-intensive electric-powered machines, which will ultimately reduce the world's reliance on unsustainable energy sources. Nowadays, many automotive companies are extremely interested in research on increased energy density of battery packs in electric vehicles so that they might become smaller, lighter and less costly. The description of activities in project P_2 is presented in Table 3.

The network diagram of the activities in project P_2 is shown in Fig. 7. The project completion time equals 33 months.

The third project (P_3) contains the development of lightweight and sustainable green materials and their application in automobile body panels. Since automotive manufacturers are aiming to make every part either recyclable or biodegradable, there still seems to be a need for development of green composites. These bio-based composites will enhance mechanical strength and acoustic performance, reduce material weight and fuel consumption, reduce production cost, improve passenger safety, and improve biodegradability for the auto parts (Ashori 2008). Green composites derived from renewable resources are bringing very promising potential to provide benefits to companies, the natural environment and end-customers, especially considering the dwindling petroleum resources (Koronis et al. 2013). The description of activities in project P_3 is presented in Table 4.

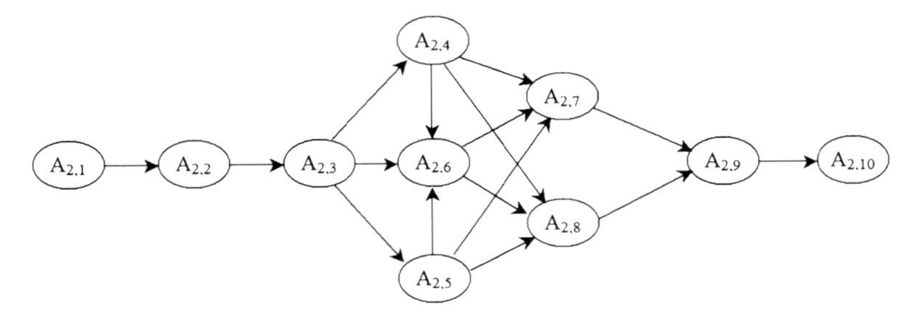

Fig. 7 Network diagram for project P_2

Table 4 Description of activities in project P_3

Activity	Description	Duration (months)	Predecessor(s)
$A_{3,1}$	Objective definition	1	–
$A_{3,2}$	Research design	1	$A_{3,1}$
$A_{3,3}$	Legal regulations, patents and environmental issues	1	$A_{3,2}$
$A_{3,4}$	Lightweight and sustainable green materials development	6	$A_{3,3}$
$A_{3,5}$	The application of green composites in automobile body panels	6	$A_{3,3}$
$A_{3,6}$	Drafting specifications and out-sourcing	4	$A_{3,4}$, $A_{3,5}$
$A_{3,7}$	Trial version of the products	1	$A_{3,4}$, $A_{3,5}$
$A_{3,8}$	Component tests and endurance test	3	$A_{3,6}$, $A_{3,7}$
$A_{3,9}$	Start of production	2	$A_{3,8}$

The network diagram of the activities in the project P_3 is shown in Fig. 8. The project completion time equals 18 months.

The above-presented projects can be described as $P = \{P_1, P_2, P_3\}$. In turn, their activities are specified as follows: $P_1 = \{A_{1,1}, \ldots, A_{1,11}\}$, $P_2 = \{A_{2,1}, \ldots, A_{2,10}\}$, $P_3 = \{A_{3,1}, \ldots, A_{3,9}\}$. It is assumed that the time horizon for the project portfolio equals 30 months ($H = \{0, 1, \ldots, 30\}$) and the number of employees is

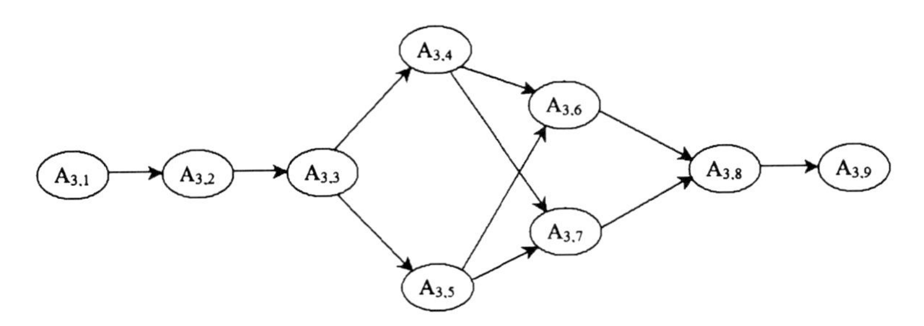

Fig. 8 Network diagram for project P_3

Table 5 Number of available employees for project activities

	A_1	A_2	A_3	A_4	A_5	A_6	A_7	A_8	A_9	A_{10}	A_{11}
$R_{1,1}$	1	1	1	3	1	1	3	3	1	1	1
$R_{1,2}$	1	1	1	3	3	3	3	1	1	1	–
$R_{1,3}$	1	1	1	3	3	1	1	1	1	–	–

limited according to the values in Table 5. The budget of project portfolio is fixed at 250 monetary units (m.u.). The sequences of activity duration for the above-presented projects are as follows (see Table 2, 3, 4): $T_1 = (2, 1, 1, 6, 2, 2, 6, 6, 1, 4, 3)$, $T_2 = (2, 2, 1, 9, 6, 3, 6, 2, 6, 4)$, $T_3 = (1, 1, 1, 6, 6, 4, 1, 3, 2)$.

The employees can conduct an activity both independently and in parallel. In the case of a few employees simultaneously carrying out an activity, the monthly cost of activity execution is as follows: 2 m.u. for one employee, 5 m.u. for two employees, and 9 m.u. for three employees. The reduction of activity duration is proportional to the number of employees. For example, if the activity's duration is planned for 6 months, and two employees are assigned to its execution, then its duration will shorten to 3 months. This leads to multicriteria optimization, where the decision-maker chooses the desired criterion.

5.2 The Direct Approach

5.2.1 Planning at the Beginning of the Project

A standard routine query formulated for the considered problem in the direct way can be stated as follows: what portfolio makespan follows from the given project constraints specified by the activity duration times, the admissible amount of resources and their allocation to project activities?

The solution to the problem results in the determination of the starting time of project portfolio activities s_{ij}, and the allocation of resources (employees) to the activities $dp_{i,j,k}$. For the above-presented project portfolio, the following sequences can be considered: $S_1 = (s_{1,1}, \ldots, s_{1,11})$, $S_2 = (s_{2,1}, \ldots, s_{2,10})$, $S_3 = (s_{3,1}, \ldots, s_{3,9})$, $Dp_1 = (dp_{1,1,1}, \ldots, dp_{1,11,1})$, $Dp_2 = (dp_{2,1,1}, \ldots, dp_{2,10,1})$, $Dp_3 = (dp_{3,1,1}, \ldots, dp_{3,9,1})$. However, the large size of the instance and the run time connected with it imposes a limitation of the solutions sought. Two cases are further considered: the first in which all activities start as soon as possible (the following sequences are sought—Dp_1, Dp_2, Dp_3), and the second in which all activities start as soon as possible with the exception of one activity (the following sequences are sought—Dp_1, Dp_2, Dp_3, $s_{2,9}$).

The reference model encompassing the assumptions of the considered example was implemented in the Oz Mozart programming environment and tested on an AMD Turion(tm) II Ultra Dual-Core M600 2.40 GHz, RAM 2 GB platform. Table 6 presents the results of solution seeking for the different strategies of

Table 6 Comparison of strategies for variable distribution

Case	Distribution strategy	Number of solutions	Depth	Time [sec]
Dp_1, Dp_2, Dp_3	Naïve	857	19	9.7
Dp_1, Dp_2, Dp_3	First-fail	857	19	9.8
Dp_1, Dp_2, Dp_3	Split	857	19	9.0
Dp_1, Dp_2, Dp_3, $s_{2,9}$	Naïve	857	32	41.4
Dp_1, Dp_2, Dp_3, $s_{2,9}$	First-fail	857	32	66.5
Dp_1, Dp_2, Dp_3, $s_{2,9}$	Split	857	23	57.6

variable distribution and two considered cases. The size of the instance equals 19,682 and 275,561 for the first (Dp_1, Dp_2, Dp_3), and the second case (Dp_1, Dp_2, Dp_3, $s_{2,9}$), respectively. The results show that the Naïve and Split distribution strategy outperforms the First-fail distribution strategy.

In the first admissible solution (for the case Dp_1, Dp_2, Dp_3—all activities start as soon as possible), the total cost equals 220 m.u., the total time—21 months, and the employee's assignment has the following form: $Dp_1 = (1, 1, 1, 1, 1, 1, 1, 2, 1, 1, 1)$, $Dp_2 = (1, 1, 1, 3, 2, 2, 3, 1, 1, 1)$, $Dp_3 = (1, 1, 1, 1, 1, 1, 1, 1, 1)$. Figure 9 presents the schedule for the considered project portfolio in which the sequences of activity starting time are as follows: $S_1 = (0, 2, 3, 4, 4, 4, 4, 10, 10, 13, 17)$, $S_2 = (0, 2, 4, 5, 5, 8, 9, 9, 11, 17)$, $S_3 = (0, 1, 2, 3, 3, 9, 9, 13, 16)$. This variant is further considered as a basic variant.

The presented approach allows the decision-maker to consider a wide range of analyses, including the trade-off analysis in a project environment. For instance, not all variants are equally profitable and suitable for further, detailed analysis. Let us assume that the decision-maker would like to get the information about a feasible duration of project portfolio within 24 months, with the total cost not greater than 207 m.u. Each variant can be assessed according to the following criterion:

$$V = (PC - BC)/(BT - PT) \tag{10}$$

where:

PC—the planned cost of a project variant,
BC—the cost of the basic variant of project,
BT—the time of the basic variant of project,
PT—the planned time of a project variant.

For the considered example, the cost of the basic variant of project portfolio equals 220 m.u. and the time of project portfolio execution equals 21 months. The feasible variants of project portfolio completion for the above-described constraints are presented in Table 7.

The obtained variants may be assessed according to the planned total cost and time of project completion, as well as the time–cost criterion (V). According to the last criterion, the most advantageous is the first variant which by the minimal cost

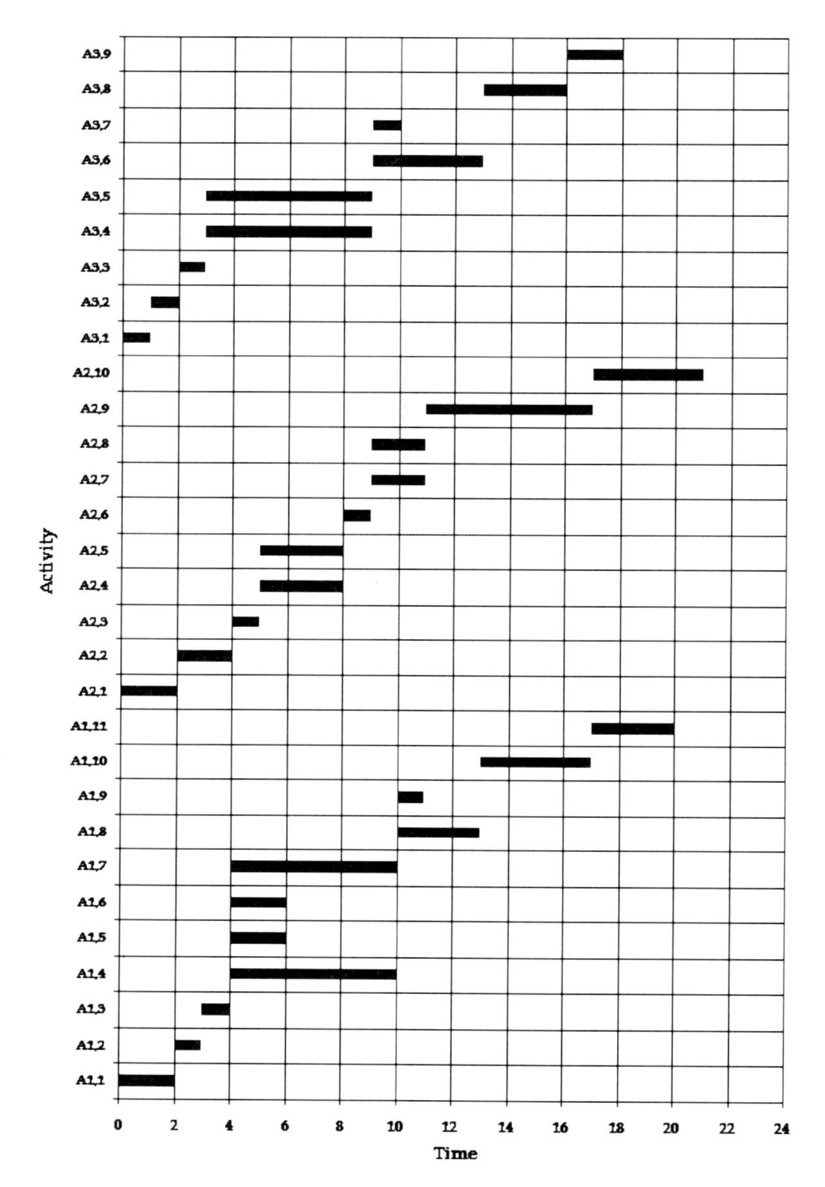

Fig. 9 Project portfolio schedule

increase allows the maximal time to be reduced. The sequences of activity starting time are as follows: $S_1 = (0, 2, 3, 4, 4, 4, 4, 10, 10, 16, 20)$, $S_2 = (0, 2, 4, 5, 5, 9, 10, 10, 13, 19)$, $S_3 = (0, 1, 2, 3, 3, 9, 9, 13, 16)$.

Table 7 Trade-off analysis of project portfolio completion

No. variant	Employee assignment	PT	PC	V
1	$Dp_1 = (1, 1, 1, 1, 1, 1, 1, 1, 1, 1, 1)$	23	204	8
	$Dp_2 = (1, 1, 1, 2, 1, 2, 2, 1, 1, 1)$	23		
	$Dp_3 = (1, 1, 1, 1, 1, 1, 1, 1, 1)$	18		
2	$Dp_1 = (1, 1, 1, 1, 1, 1, 1, 1, 1, 1, 1)$	23	207	6.5
	$Dp_2 = (1, 1, 1, 2, 1, 2, 2, 1, 1, 1)$	23		
	$Dp_3 = (1, 1, 1, 1, 2, 1, 1, 1, 1)$	18		
3	$Dp_1 = (1, 1, 1, 1, 1, 1, 1, 1, 1, 1, 1)$	23	207	6.5
	$Dp_2 = (1, 1, 1, 2, 1, 2, 2, 1, 1, 1)$	23		
	$Dp_3 = (1, 1, 1, 2, 1, 1, 1, 1, 1)$	15		
4	$Dp_1 = (1, 1, 1, 1, 1, 1, 1, 1, 1, 1, 1)$	23	207	6.5
	$Dp_2 = (1, 1, 1, 2, 1, 2, 3, 1, 1, 1)$	22		
	$Dp_3 = (1, 1, 1, 1, 1, 1, 1, 1, 1)$	18		
5	$Dp_1 = (1, 1, 1, 1, 1, 1, 1, 1, 1, 1, 1)$	23	207	6.5
	$Dp_2 = (1, 1, 1, 2, 2, 2, 2, 1, 1, 1)$	23		
	$Dp_3 = (1, 1, 1, 1, 1, 1, 1, 1, 1)$	18		
6	$Dp_1 = (1, 1, 1, 1, 1, 1, 1, 2, 1, 1, 1)$	20	207	6.5
	$Dp_2 = (1, 1, 1, 2, 1, 2, 2, 1, 1, 1)$	23		
	$Dp_3 = (1, 1, 1, 1, 1, 1, 1, 1, 1)$	18		
7	$Dp_1 = (1, 1, 1, 1, 1, 1, 2, 1, 1, 1, 1)$	20	207	6.5
	$Dp_2 = (1, 1, 1, 2, 1, 2, 2, 1, 1, 1)$	23		
	$Dp_3 = (1, 1, 1, 1, 1, 1, 1, 1, 1)$	18		
8	$Dp_1 = (1, 1, 1, 2, 1, 1, 1, 1, 1, 1, 1)$	23	207	6.5
	$Dp_2 = (1, 1, 1, 2, 1, 2, 2, 1, 1, 1)$	23		
	$Dp_3 = (1, 1, 1, 1, 1, 1, 1, 1, 1)$	18		

5.2.2 Re-Planning of the Project Portfolio

The project portfolio review at the 6th month indicates that the first and third projects are conducted according to plan. In turn, in the second project, the activity $A_{2,1}$ was completed within 3 months, the activity $A_{2,2}$ within 2 months, and the activity $A_{2,3}$ within 1 month. Re-estimation indicates that the duration of activities in the project P_2 takes the following sequence: $T_2 = (3, 2, 1, 12, 6, 3, 6, 2, 6, 4)$. The admissible assignment of employees to the activities reduces the predicted time of project completion to 23 months (the predicted cost equals 225 m.u.), however, this still exceeds the project deadline (21 months).

In the considered case, the set of admissible solutions is empty, i.e. there is no schedule that fulfils all assumed constraints. Thus, there is still a possibility to reformulate the considered problem by stating it in an inverse approach, i.e. the approach aimed at searching for decision variables (e.g. reallocation of resources) ensuring that the completion time of the considered project will not exceed the assumed deadline H. Such a case is considered in the next subsection.

Table 8 Feasible variants for the inverse approach

No. variant	Employee assignment	*PT*	*PC*	*V*
1	$Dp_1 = (1, 1, 1, 1, 1, 1, 2, 1, 1, 1, 1)$	20	230	2.5
	$Dp_2 = (1, 1, 1, 3, 1, 2, 3, 1, 1, 2)$	21		
	$Dp_3 = (1, 1, 1, 1, 1, 1, 1, 1, 1)$	18		
...	...	...	...	...
48	$Dp_1 = (1, 1, 1, 1, 1, 1, 2, 1, 1, 1, 1)$	20	243	6
	$Dp_2 = (1, 1, 1, 3, 3, 3, 3, 1, 2, 2)$	18		
	$Dp_3 = (1, 1, 1, 1, 1, 1, 1, 1, 1)$	18		

5.3 The Inverse Approach

Similarly as in the previous subsection, for the direct approach, the same activity networks and the constraints concerning the deadline and budget of the project portfolio are considered. The problem can be stated as follows: What amount of resources and their allocation ensures that the project portfolio does not exceed the deadline and the budget? The response to this question is connected with the determination of sequences concerning the number of employees assigned to the projects (*Dp*) and starting time of activity (*S*).

Let us assume that a decision-maker considers the assignment of an additional employee to one of the last two activities in the project P_2 to verify the possibility of the fulfilment of the above described constraints. There are 48 feasible variants of the employee's assignment to the project activities, and the solution has been obtained in less than 1 s. The first and last admissible solutions of project portfolio completion are presented in Table 8. To determine the time–cost criterion (V), the values from Sect. 5.2.2 are retrieved, i.e. the total cost in basic variant equals 225 m.u. and the total time for project portfolio equals 23 months.

According to the time–cost criterion, the most advantageous is the first variant that by the minimal cost increase allows the maximal time to be reduced. The obtained solution provides a plan for project portfolio execution as well as a base for further adjustment aimed at fitting to real live execution.

6 Conclusions

Projects are becoming a more and more important concern in the activities of present day companies. This trend especially occurs in organizations that are based on multi-product management, as in an automotive company. Hence, there is an increase in demand for new knowledge that enables the solution of problems encountered during complex project portfolio execution. In this case, knowledge concerning project management, especially the rescue of failed projects, is particularly significant.

The efficient development of a task-oriented decision support system implies the formulation of a single knowledge base that results in a reference model of project planning. This model, in terms of constraint satisfaction problems can be described as including the sets of decision variables, their domains, and constraints which link and limit these variables. The hierarchical and open structure of model enables the solution of decision problems with varying levels of specificity. The decision problems can contain a query about the results of proposed decisions as well as decisions ensuring the expected results.

The limitations of existing commercially available tools (lack of abilities to solve problems defined in the inverse approach, lack of support new automotive product development in terms of CSP, etc.) was the motivation to develop a design methodology for task oriented decision support systems aimed at automotive portfolio planning. The proposed approach assumes an open structure reference model that takes into account different types of variables and constraints as well as permits to formulate direct and inverse project planning problems. The model supports descriptive statement of the problem followed by its implementation in one of the constraint logic languages.

The advantage of the proposed approach is at least twofold. Firstly, there is the description of an automotive company and project management in terms of a single knowledge base. This facilitates the use of constraint programming to develop a decision support system. Secondly, the decision support system seeks a set of feasible variants for project completion that do not exceed constraints, e.g. budget, deadline, amount of resources, as well as it assesses these variants according to assumed criteria. This is especially attractive in the case of a lack of a possibility for continuing the baseline schedule and supports managers in making decisions concerning the completion or abandonment of a project at risk of failure. From this point of view, the proposed approach is an extension to a standard ERP system. It can partially retrieve data from an ERP system concerning, for instance, the amount of resources (employees, financial means) and their accessibility in the given time.

The proposed approach enables a set of schedules concerning alternative project portfolio completion to be obtained and assessed. It results in reducing project portfolio execution and its cost, as well as accelerating the launch of new car models containing more environmentally friendly technologies. As a consequence, it increases the effectiveness of project portfolio management, and the competitiveness of an automotive company, thus providing benefits to the companies and customers, as well as the natural environment.

The hypothesis made in critical path method that activity durations are deterministic and known is rarely satisfied in real life (Maravas and Pantouvakis 2012). Therefore, further research focuses on the formulation of the reference model in terms of distinct and imprecise variables. The proposed model will be also extended to other fields of project management, including project risk. Moreover, further research may be aimed at developing task-oriented searching strategies, the implementation of which could interface a decision-maker with a user-friendly intelligent support system. Thus, the decision support tool could be more easily accepted by a user, enabling real-life verification.

Acknowledgments The work was supported by the National Scholarship Programme of the Slovak Republic.

References

Asholi A (2008) Wood-plastic composites as promising green-composites for automotive industries! Bioresour Technol 99:4661–4667
Badra F, Servant F-P, Passant A (2011) A semantic web representation of a product range specification based on constraint satisfaction problem in the automotive industry. In: Proceedings of the 1st international workshop on ontology and semantic web for manufacturing, pp 37–50
Banaszak Z, Bocewicz G (2011) Decision support driver models and algorithms of artificial intelligence. Warsaw University of Technology, Faculty of Management, Warsaw
Bocewicz G, Banaszak Z, Muszyński W (2009) Decision support tool for resource allocation subject to imprecise data constraints. In: IEEE international conference on control and automation, pp 1217–1222
Bonnal P, Gaurc K, Lacoste G (2004) Where do we stand with fuzzy project scheduling? J Construct Eng Manage 130(1):114–123
Brailsford SC, Chris N, Potts ChN, Smith BM (1999) Constraint satisfaction problems: algorithms and applications. Eur J Oper Res 119:557–581
Burke EK, Kendall G (eds) (2005) Search methodologies. Springer, New York
Chanas S, Komburowski J (1981) The use of fuzzy variables in PERT. Fuzzy Sets Syst 5(1):11–19
Dincbas M, Simonis H, Van Hentenryck P (1988) Solving the car sequencing problem in constraint logic programming. In: Proceedings of the European conference on artificial intelligence, Munich
Dubois D, Fortin J, Zieliński P (2010) Interval PERT and its fuzzy extension. In: Studies in fuzziness and soft computing, vol 252. Springer, pp 171–199
Fortin J, Zielinski P, Dubois D, Fargier F (2010) Criticality analysis of activity networks under interval uncertainty. J Sched 13(6):609–627
Grimm K (2003) Software technology in an automotive company: major challenges. In: Proceedings of the 25th international conference on software engineering, pp. 498–503
Jüngen FJ, Kowalczyk W (1995) An intelligent interactive project management support system. Eur J Oper Res 84:60–81
Kerzner H (2009) Project management: a systems approach to planning, scheduling, and controlling, 10th edn. John Wiley and Sons, New York
Koronis G, Silva A, Fontul M (2013) Green composites: a review of adequate materials for automotive applications. Compos B Eng 44:120–127
Levis HF, Slotnick SA (2002) Multi-period job selection: planning works loads to maximize profit. Comput Oper Res 29:1081–1098
Maravas A, Pantouvakis JP (2012) Project cash flow analysis in the presence of uncertainty in activity duration and cost. Int J Project Manage 30:374–384
Mayyas A, Qattawi A, Omar M, Shan D (2012) Design for sustainability in automotive industry: a comprehensive review. Renew Sustain Energy Rev 16:1845–1862
Müller D, Herbst J, Hammori M, Reichert M (2006) IT support for release management processes in the automotive industry. In: Dustdar S et al. (eds) BPM. Lecture notes in computer science, vol 4102. Springer, Heidelberg, pp 368–377
Pesic M, van der Aalst WMP (2006) A declarative approach for flexible business processes management. In: Business process management workshops. Lecture notes in computer science, vol. 4103, Springer, pp 169–180

Power DJ, Sharda R (2007) Model-driven decision support systems: concepts and research directions. Decis Support Syst 43:1044–1061

Project Management Institute (2008) A guide to the project management body of knowledge, 4th edn. Paperback PMI, PMBOK Books, Newton Square

Relich M (2011) Project prototyping with application of CP-based approach. Management 15(2):364–377

Rossi F, van Beek P, Walsh T (2006) Handbook of constraint programming. Elsevier Science, Amsterdam

Sawik T (2007) Multi-objective master production scheduling in make-to-order manufacturing. Inter J Oper Res 45(12):2629–2653

Turner R (2006) Towards a theory of project management: the functions of project management. Int J Project Manage 24:187–189

Turner R, Ledwith A, Kelly J (2010) Project management in small to medium-sized enterprises: matching processes to the nature of the project. Int J Project Manage 28:744–755

Van Roy P, Haridi S (2004) Concepts, techniques and models of computer programming. Massachusetts Institute of Technology, Cambridge

What is Influencing the Sustainable Attitude of the Automobile Industry?

Angel Peiró-Signes, Ana Payá-Martínez, María-del-Val Segarra-Oña and María de-Miguel-Molina

Abstract The automotive industry is considered one of the most environmental aware manufacturing sectors. The aim of this study is to determine if there are any characteristics that differentiate automotive companies in environmental attitudes through the quantitative analysis of a sample of 224 companies belonging to the Spanish automobile industry. The chapter also provides an overview of the implications and constrains due to the integration of eco-design in a real company of the automotive sector giving a vision of the legislation applicable referred to eco-design, the adaptation of the constructor standards and some examples of general rules of the product and process design to satisfy these requirements, as well as the environmental policy that affects its decisions.

Keywords Automotive industry · Sustainability · Eco-design · Process and product design

A. Peiró-Signes (✉) · M. Segarra-Oña · M. de-Miguel-Molina
Universitat Politècnica de València, Valencia, Spain
e-mail: anpeisig@omp.upv.es

M. Segarra-Oña
e-mail: maseo@omp.upv.es

M. de-Miguel-Molina
e-mail: mademi@omp.upv.es

A. Payá-Martínez
Faurecia Certification and Systems, Valencia, Spain
e-mail: ana.paya@faurecia.com

P. Golinska (ed.), *Environmental Issues in Automotive Industry*,
EcoProduction, DOI: 10.1007/978-3-642-23837-6_3,

1 Introduction

The automotive industry is directing its efforts to incorporate the principle of the sustainable development, considering the service life of the product and involving the suppliers from first stage of design.

Suppliers play a vital role in the development and production of the vehicle; therefore they must be aligned with the customer environmental requirements and any applicable legislation. These requirements are provided by the customer at the beginning of the project and the requirements will be different depending on the commodity.

In order to be able to demonstrate the possibility of reaching the ratio of recycling and the absence of dangerous substances that allows the vehicle homologation, the vehicles constructors include in their requirements the report of all the materials and substances remaining on a vehicle at the sale point via the IMDS system (International Material Data System).

In the case of Ford Motor Company, for example, the certification according to ISO 14001 is an indispensable requirement to work as a supplier and the report in IMDS is a part of the Production Part Approval Process, which is the process for vehicle components' homologation. As an evidence of the initiative of the sector, it is possible to even emphasize that before the European Union adopted REACH (Registration, Evaluation, Authorization and Restriction of Chemical substances) regulation on chemical management or on end-of-life vehicles, companies like Ford Motor Company already had a material management system that allowed them to track all the substances and materials of the vehicle.

Another way to assure the fulfilment of the regulation is including this requirement as a part of the Manufacturing Site Assessment, that is, including this concept as a key element of the Q1, a set of quality and production disciplines and some indicators that Ford suppliers must follow to allow suppliers measurement of client expectations fulfilment.

Aligned with directive 200/53/EC, Ford Motor Company demands its suppliers that all the pieces must go labelled according to company standards, for later better recycling.

Suppliers must provide evidences of Design-for-Environment principles implementation (DfE), including Design-for-Disassembling (DfD) and Design-for-Recycling (DfR). One of the tools provided by the manufacturer of vehicles is the Design Verification Method to verify components recyclability by a recyclability evaluation.

The increase of recycled materials use is a requirement of the European directive and Ford Company requires suppliers to incorporate recycled materials in suitable components.

With the purpose of guaranteeing the requirement to provide the necessary information to the recycling companies so that end-of life vehicles are eliminated in a safe, economic and respectful way with the environment, several manufacturers of vehicles joined to create IDIS database (International Dismantling Information System), which compiles all this information.

During long time, the indicators considered as "classical" used by the companies in order to establish their targets have been dealing only with economic data, without considering social or environmental aspects.

To make sustainability a reality, the existing measuring tools need to determine where we are now and how far we need to go and whether humanity's demand remains within the interests of the globe's natural capital stocks (Wackernagel et al. 1999).

The ecological footprint indicator tries to fill in the gap. The ecological footprint is a measure of the load imposed by a given population on nature. It represents the land area necessary to sustain current level of resource consumption and waste discharge by that population (Wackernagel 2002).

By measuring the footprint of a population we can assess our pressure on the planet and will allow us to know how much nature we have, how much we use, and to track our progress toward the goal of a sustainable development (Munksgaard et al. 2005; Jorgenson and Burns 2007).

Today humanity uses the equivalent of 1.5 planets to provide the resources we use and absorb our waste. This means, it now takes the Earth 1 year and 6 months to regenerate what we use in a year.

In order to amend this situation it is important to establish sustainable development strategies. Implementing environmental strategies in the design phase will allow reducing the environmental impact of products during their service life and at their end of life (Wimmer et al. 2010) as in other leading industries (Criado 2007; Bohdanowicz 2005; Miret et al. 2011).

The consumer is increasingly becoming a citizen above all else, considering the ecological footprint of the products he or she buys, and the environmental impact of products. The question of use has become essential. When a choice is available, consumers are increasingly weighing their personal needs against their responsibility as citizens (Clark et al. 2009).

Some developed solutions to improve product design allowing its later recycling and reusability, although, at the moment, there are certain limitations in the technologies available in each country as many countries have still not invested in any specific treatment installation.

2 The Automotive Industry and the Environmental Awareness

As a result of these evolutions, the automobile status and role in daily life are changing. The automobile is becoming one choice among many in daily mobility, along with other means of individual transport (bicycles, scooters, rental cars) and collective transport (mass transit, carpooling, trains), although still indispensable in many areas around the world (Whitmarsha and Köhlerb 2010).

In this sense, the automotive industry, like other industries, must face the challenge to adapt product and process design to integrate environmental aspects (Schiavonne et al. 2008) as, every year, end-of life vehicles (ELVs) in the UE generate between eight and nine million tonnes of waste, which must be managed correctly (European Commission 2000). Volume of ELVs arising each year is increasing and it is expected to be 14 million tonnes by 2015 (Eurostat 2011). In this sense, Spain has to play a crucial role as five countries (Germany, UK, France, Spain and Italy) are responsible for approximately 75 % of EU 25 vehicle de-registrations (Eurostat 2011) and also holds an important part of Europe's automotive industry.

The present legislation tries to promote the use of recycled materials in the development of new vehicles and this is affecting the innovation processes of the companies, but is this compatible with customer specifications? How to use recycled materials in components that need to fulfil some certain mechanical or aesthetic characteristics? (Gerrard and Kandlikar 2007).

Environmental impact reduction and the consideration of the complete service life of the product must be objectives to deal with engineering specifications and to provide solutions with similar production costs as traditional development to date. Thus, design of vehicles for recycling and recovery, including their components and materials, as spare and replacement parts, might contribute to the protection, preservation and improvement of the environment quality and to energy conservation, although, waste generation must be avoided as much as possible (Santini et al. 2010; Ferrão and Amaral 2006).

Although, systems and facilities requirements for the collection, treatment and recovery of end-of life vehicles are important to ensure the attainment of the targets for reuse, recycling and recovery, producers meet all, or a significant part of the costs of all measures taken, and end-of life vehicle owner should deliver the vehicle into an authorized treatment facility without any cost. Then, design phase becomes crucial to ensure also companies profitability (Orsato and Wells 2007).

Preventive measures applied from the conception phase of the vehicle will improve the reduction and control of hazardous substances in vehicles, in order to prevent their release into the environment, and will facilitate recycling and avoid the disposal of hazardous waste.

Appropriate design will also help to ensure that certain materials and components do not become shredder residues, and are not incinerated or disposed of in landfills.

This will end integrating requirements for dismantling, reuse and recycling of end-of life vehicles and their components in the design and production of new vehicles.

The constant increase of the quantity for reuse, recycling and recovery must be a target in producers design phase as environmental policies pressures in that way and consumers are increasingly better informed and adjust their behaviour and attitudes towards environmental friendly products. Companies have been leading the efforts to incorporate the principles of designing for sustainability and the use of a lifecycle management approach (Segarra et al. 2011a).

Design for Environment principles into the product development process started in the early 1990s, however, they were initially focused on designing vehicles to facilitate end-of-life disassembly and recycling by taking into account the accessibility of parts to be disassembled, the type and number of different fasteners used and the marking of parts for easy identification.

Based on several studies, it became clear that focusing on a single lifecycle phase (e.g., end of life) leads to sub-optimizations and potentially increased impacts in other lifecycle phases.

Since then, companies have shifted their focus to include a more comprehensive lifecycle approach to improve the sustainability of vehicles, by incorporating the material and component production and the use phases, in addition to the end-of-life phase (Leduc et al. 2010).

Also, sustainability management tools have been applied in the new vehicles. These tools incorporate societal and economic aspects as well as environmental aspects into lifecycle analysis and design approach.

3 Regulation

The increase of the ecological awareness has been supported by the introduction of environmental regulation. The vehicles at the end of their life generate a great amount of residues that should have to be reused and recycled. The achievement of this activity is predefined especially due to the Directive 2000/53/EC, established by the European Parliament and the Council of End of Life Vehicles (ELV). The directive is been transferred to the different national laws and its main objective is the prevention of waste from vehicles and, in addition to this, the reuse, recycling and other forms of recovery of end-of-life vehicles and their components so as to reduce the disposal of waste. The Directive also aims to improve the environmental performance of all economic operators involved in the life-cycle of vehicles.

In order to be able to meet these objectives, the European Union has unfolded new requirements for vehicle manufacturers assuring the design of recyclable vehicles.

Priority must be given to the reuse and recovery (recycling, regeneration, etc.) of vehicle components aiming to increase its rate.

The reuse and recovery rate (in average weight per vehicle and year) should reach 85 % no later than January 1st 2006 and 95 % no later than January 1st 2015 (see Fig. 1).

The reuse and recycling rate (in average weight per vehicle per year) should reach 80 % no later than January 1st 2006 and 85 % no later than January 1st 2015 (see Fig. 1).

By the fulfilment of this directive it is expected to reduce the production of residues by limiting the use of hazardous substances in new vehicles, by designing

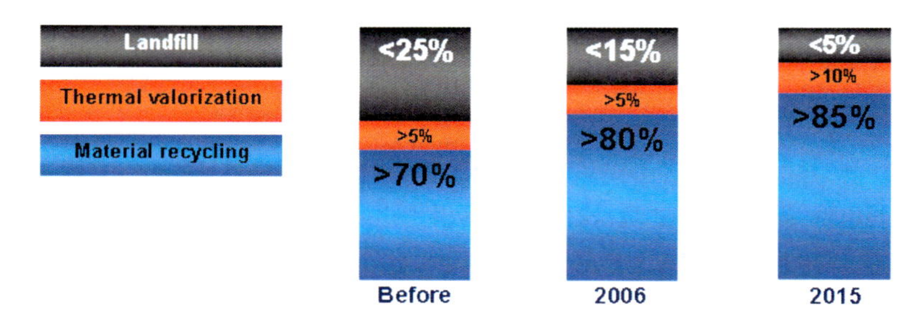

Fig. 1 Reuse and recycling rates

and producing vehicles which facilitate reuse and recycling and by developing the integration of recycled materials.

In order to make dismantling easier, manufacturers should meet codification standards for materials in their components to allow their identification, and should provide the necessary information to be able to realize the disassembling of the components (Fig. 2).

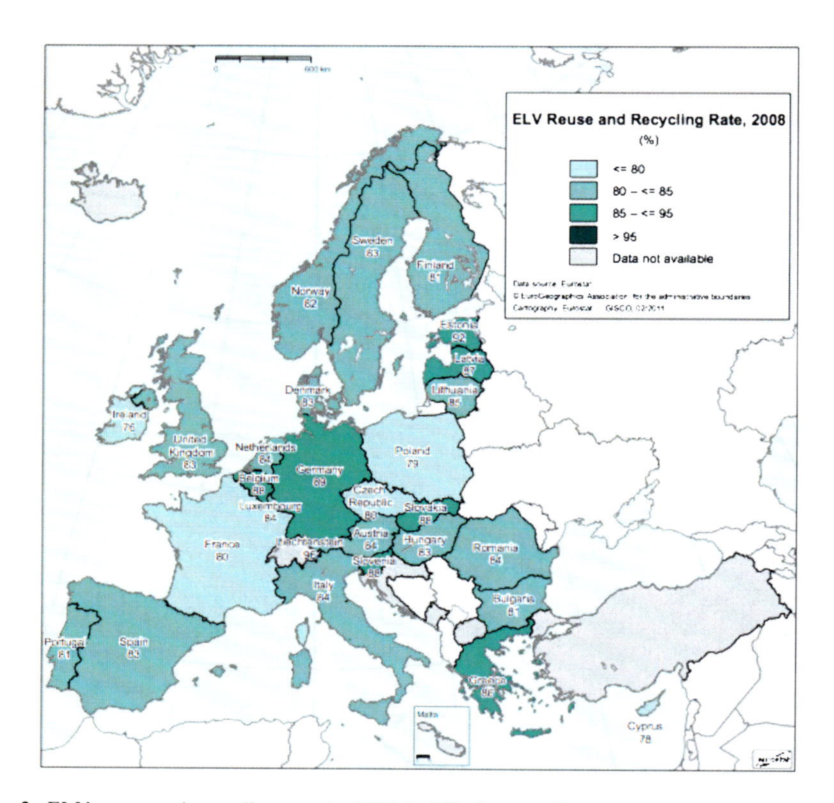

Fig. 2 ELV reuse and recycling rate in 2008 in EU, Source: Eurostat

Moreover, vehicles may be put on the market only if they meet the provisions of the EN ISO 22628:2002 (Road vehicles—Recyclability and recoverability—Calculation method).

As it is shown, government policies in developed countries have encouraged the development of markets for recycled materials, a certificate of destruction for the de-registration of end-of life vehicles, the growth of collection and treatment operators while recyclability and recoverability of vehicles have been promoted.

But, it is important to lay down requirements not only for storage and treatment operations but for vehicle and components manufacturing world round in order to avoid the emergence of distortions in trade and competition, especially with the developing countries.

4 Research Methodology

In this study, first, a collection and interpretation of data related to eco-design and sustainable development in the automotive sector has been made in order to characterize environmental orientation of the automotive industry companies while innovating.

Like most industries, the automotive industry is confronted with environmental issues namely: vehicle emissions, non-renewable material and energy consumption, generation of waste during production and at the end of life. There is an increasing stakeholders awareness, which finds its expression in new regulation and customer requirements (Manzini and Vezzoli 1998) Due to the wide range of environment impacts of the automotive industry different strategies have been launched to rectify this situation.

The study also tries to figure out if some of these qualitatively identified characteristics detected in leader companies of the automotive industry are followed by others and quantitative research will try to identify a model to differentiate environmental proactive companies from not proactive based on key indicators that have been identified in other studies (Segarra et al. 2011b).

To do so, a quantitative analysis has been performed. To establish environmental proactivity, four variables (used as dependent in each model) were taken into account:

- Importance of less materials per unit produced while innovating
- Importance of less energy consumption per unit produced while innovating
- Importance of less environmental impact while innovating
- Importance of environmental legislation requirements while innovating

It is important to identify the characteristics of the firms under study since these characteristics affect their environmental management practices.

This research analyzes a sample of 224 companies belonging to the Spanish automobile industry, some of them belonging to multinational companies. In line with Dubé and Pare (2003), well-known standardized statistical analysis methods, such as analysis of variance and regression analysis, have helped researchers confirm or reject hypotheses in quantitative research.

To reduce data variables a factorial analysis method was applied, allowing us to obtain homogeneous correlated variable groups. Furthermore, a logistic regression model, with the previous factor analysis results, was made to fit data.

The data was collected from PITEC database (Technological Innovation Panel), which consists of a statistical tool to monitor the technological innovation activities of Spanish companies. The database was built by the INE (Spanish National Statistics Institute) with the advice of academics and experts. A total of 255 variables were analyzed including a comprehensive list of Spanish companies which are characterized by the type of innovation (classified by the Oslo Manual, 2005) that they undertake, by industry (in line with the Spanish National Activities Classification, CNAE) or by geographical location. Data from 2009, the latest data available, was used for the analysis.

Variables included in this study were selected according to theoretical statements. Net sales (NS) represents the total sales income, size by number of employees (SZ) represents the number of full-time employees in the company, total goods investment (INVER) represents gross investment in tangible goods.

National market (MDONAC) indicates whether the companies operate on a national scale. E.U. market (MDOUE) indicates whether the companies operate on a European scale. Worldwide market (OTROPAIS) indicates whether the companies operate on a world wide scale rather than European Union. These are binary variables with $1 = $ Yes and $0 = $ No. The number of total patents (PATNUM), the number of European patents (PATEPO) and the number of national patents (PATOEPM) were measured respectively as the number of patent applications and the patents at the European and the Spanish levels.

Total investment in R&D activities (GTINN) represents the total expenditure in internal and external R&D activities and number of R&D employees (PIDCA) represents the number of full-time employees who work on R and D activities.

4.1 *Qualitative Approach*

4.1.1 Environmental Awareness at the Automobile Supplier Level: Faurecia's Case

Companies that work as vehicle manufacturers suppliers are a key element in a design that reduces the environmental impact of the automobile.

Faurecia is a specialist in the engineering and production of automotive components holds global leadership status in each of its core businesses: Automotive Seating, Emissions Control Technologies, Interior Systems and Automotive Exteriors. Its customer portfolio features practically every automaker around the world, including manufacturers in emerging economies, such as the Indian, Chinese and Korean markets.

With the aim of preparing good technical and economic practices, as well as fulfilling the effective legislation and the requirements of client, Faurecia has a set of specialists who form the eco-design department. This department is the one in charge of developing work standards that assure environmental aspects integration in the product and process design considering the complete service life of the product.

From the use of "green materials" to recycling, from the reduction of emissions to the reduction of weight, the concept of "clean car" is a complex subject with a broad scope.

Beginning with company's ethical code, Faurecia Group undertakes to implement actions aimed at respecting the environment and improving its protection.

In carrying out their daily activities, all Faurecia employees should be aware of their responsibilities towards protecting the environment, especially through the following commitments:

- Reduce waste and polluting products, conserve natural resources and recycle materials at each step in the manufacturing process;
- Actively pursue a development policy and implement technology capable of reducing polluting emissions.
- Constantly assess the impact of its products and the activity of its plants on the immediate environment and communities with a view to making constant improvements.

Different action lines like usage footprint reduction, with a strategy of innovation centred on light-weighting and emissions control technologies, and vehicle's ecological footprint reduction from production to end of life, through the use of more environmentally friendly materials and the implementation of cleaner production processes, show the commitment of the company with the environment.

To reduce economic dependence on petroleum materials as well as the environmental impact of their products, automotive manufacturers and suppliers are including more and more biomaterials (wood, hemps, linseed, wheat, beets, etc.) in their designs. Since the 1990s the Faurecia Group has been working on technologies combining polymers and natural fibbers for door panels. These natural fibbers can make up 50–90 % of a door panel. Also, polyolefins and fibers such as hemp and sisal are used for the creation of semi-structural parts, such as bumper supports, in place of glass fibers. The objective is to demonstrate the possible substitution of petroleum-based plastics by natural bio-based materials.

Additionally, production plants use environmental management tools to reduce the negative effects of industrial activity in the environment by the application of good practices in energy saving and waste generation, promoting recycling and reusing production processes that are more kind with environment conservation.

Certification of industrial sites (environmental management systems based on the ISO 14001 international standard) and training of employees to respect the environment have progressively been implemented, so 80 % of production sites were certified and more than 50 % of the people were trained by the end of 2009.

In 2008, all of the Group's facilities achieved a 3 % reduction in overall water consumption compared to 2007 and 46 % of the water consumed was recycled internally or disposed of naturally, with the rest directed to collective treatment facilities.

Similarly, energy consumption was reduced by 1 % between 2007 and 2008. This can be broken down into 34 % natural gas, 60 % electricity, and 4 % LPG (Liquid Petroleum Gas) fuels, and less than 2 % steam.

Logistic improvements, eco-friendly non-pollutant paints usage and other action have been manifested effective in reducing emissions.

At the same time, the updated Law on New Economic Regulation introduces mandatory HSE (health safety and environment) reporting for French corporations. Faurecia as French company listed on the French stock exchange is requested by the French law to report about the social and environmental impacts of its worldwide activities, such as energy consumption, atmospheric emissions, wastes, compliance with environmental regulation and legislation.

The NER is without a doubt one of the most important sustainability milestones in Europe or North America to date. For the first time on record, all listed corporations will have to publicly report on their triple bottom line -financial social and environmental- activities in both their annual and financial reports (2003, Eva Hoffmann. Environmental Reporting and Sustainability Reporting in Europe)

In the context, where some standards are getting more and more strict on emissions of CO_2 and antipollution, manufacturers and suppliers are increasing their efforts in the control and reduction of emissions to the atmosphere. For some years, the reduction of the weight has become an automotive industry priority.

Since Faurecia's products account for 15–20 % of a vehicle's weight, a responsibility and a commitment to lighten vehicles have been made to improve mileage and reduce pollution. Faurecia innovations will allow reducing the weight of their products by as much as 30 % by 2020.

Through light-weighting, new applications of natural materials and lowering emissions, Faurecia contributes to make the world less dependent on volatile oil supplies and helps promote a cleaner environment, as the reduction of the weight of the car in 10 kg is translated in a reduction of CO2 emissions of 1 gram by kilometre.

4.1.2 Design for Recycling (Environmental Orientation While Innovating)

Virtually all of the material in today's automobiles can technically be recycled. The challenge facing engineers is making this recycling process economical, especially for materials in such components as seats and instrument panels. Recycling these components requires different materials to be separated so that each can be recycled individually (Coulter et al. 1998).

Although the European Directive is very detailed concerning automotive materials and recycling recommendations it has no detailed instructions about how to do it at designing and production level. (De Medina 2006.

The goal of the Design-for-Recycling teams is to reduce the impact of substances originated materials of end of life vehicles and to promote later reusability.

Different measures have been developed to increase the potential recycling of the products, like the recycling production waste during the production process.

Design for recycling is oriented to facilitate the same in the materials and components of end of life vehicles, that's why the technologies of recycling available must be considered. The technologies available right now are: collection and de-pollution, dismantling and shredding.

For Faurecia there are three main options of recycling. According to these three options, general limitations of design and basic rules for the evaluation of recycling are considered from the design phase.

- Dismantling: in order to identify which pieces to disassemble with the correct labelling in visible zones, diminution of the number of materials, fixations that allow fast assembly and disassembling of components
- Post shredding floatation: The strategy of this process consists in recovering PP and PE with a density less than one, so that it floats. The objective is to take advantage of this technology to use PP and EP with a maximum of 10 % of mineral additives.
- Post shredding "Bulk recycling". It consists in classifying the residues in three categories: heavy, light and mineral.

The optimal design would happen by optimizing the product being recyclable in each of the three routes mentioned.

In the drawn map next, recycling technologies available in Europe are mapped (Fig. 3)

4.2 Quantitative Analysis

An exploratory factor analysis was performed on all independent variables, using Varimax method in an attempt to understand the factor structure and the corresponding measurement quality. The solution shows three factors which account for 78.96 % of the variance and significance 0.000, namely size, open market orientation, formal innovative activity (Table 1 shows the factor analysis results). All statistical analyses were carried out using SPSS for Windows, version 18.0.

Barlett's test of sphericity was calculated with the Kaiser–Meyer–Olkin statistic, to verify the suitability of the analysis. In line with Hair et al. (1998), it is usual to accept a solution explaining over 60 % of variance in social sciences. Factor estimates as well as the assessment of the overall fit were carried out using a principal component analysis, which was suitable to summarize the original information in factors for prospective purposes (Hair et al. 1998).

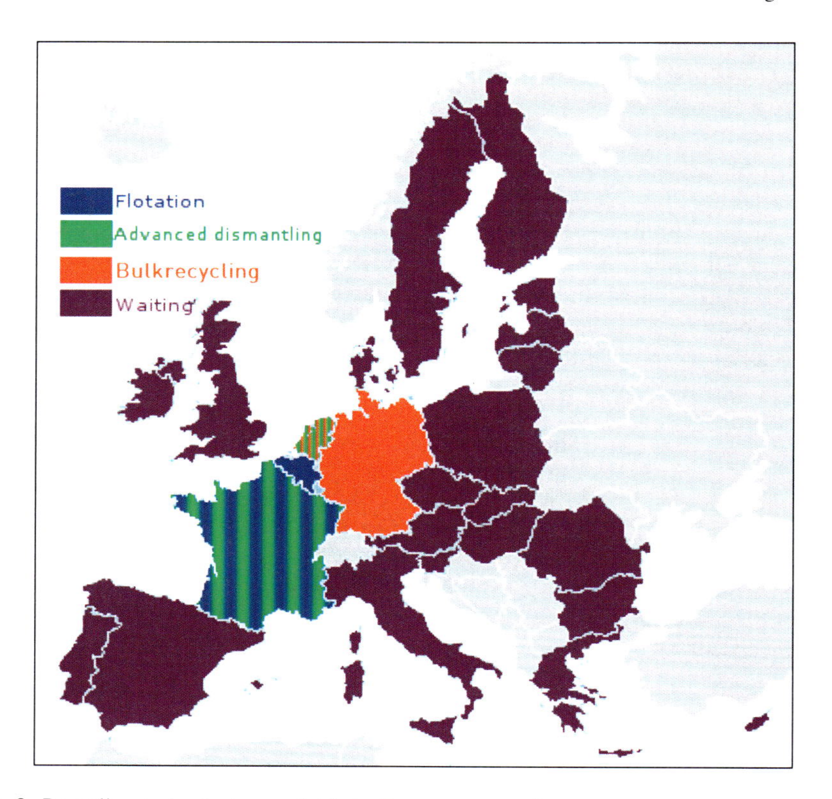

Fig. 3 Recycling technologies available in Europe per member

Table 1 Rotated component matrix

	Factor 1	Factor 2	Factor 3
NS	0.952		
INVER	0.887		
SZ	0.958		
MDONAC			0.672
MDOUE			0.821
OTROPAIS			0.773
GTINN	0.795		
PIDCA	0.817		
PATNUM		0.955	
PATOEPM		0.889	
PATEPO		0.915	

Rotation converged in four iterations. * Principal Component analysis. Varimax with Kaiser Normalization. 78.963 % variance explained -KMO, 756- Sig. 000

The factor loadings are the correlation coefficients between the variables and factors and the squared factor loading represents the percent of variance in that indicator variable explained by the factor.

High loadings have been considered to be six or higher (Hair et al. 1998) and are used to determine the cut-off.

Also scores were extracted to be used as variables in subsequent logistic regression.

The results of the Varimax rotation reinforced the expected pattern as Net sales, number of employees, total investment, total expenditure and employees in R&D activities refer to the size of the company. According to theory, larger structures involve more employees and higher sales. Moreover, the larger the company is, the greater the innovation investment (Churchill and Lewis 1987; Becker et al. 2005; Greiner 1997). All the variables which made up the first components were positively correlated and fitted the evolutionary theory of Nelson and Winter (2002).

National market (MDONAC), European Union market (MDOUE) and Worldwide market refers to market size, and thus market orientation becomes the second factor. According to theory (Salomon and Shaver Salomon 2005), innovation plays a crucial role in export behaviour and acts as a moderating factor in open market oriented firms.

Economic theory views patents as instruments aimed at fostering innovation and diffusion (Encaoua et al. 2006). The empirical evidence suggests that patents provide a fairly reliable measure of innovative activity (Acs et al. 2002) since innovation, growth and competitiveness are correlated (Crosby 2007). Thus, formal innovative activity is reflected as the third factor.

Factor scores from factor analysis were used as covariates in multinomial logistic regression. They have a mean of zero and a standard deviation of one.

The dependent variables of the models were modified from those called in PITEC database as OBJET9, OBJET10, OBJET11 and OBJET 13 which measure how essential it is for innovating firms to improve material consumption, energy consumption, the environmental impact or the environmental legislation accomplishment.

Dependent variables were recoded into binary dependent variable in order to differentiate High/medium oriented (value = 1) from Low/Not oriented firms (Value = 0) so they were designated with the suffix MOD.

Logistic regression can be used to predict a categorical dependent variable on the basis of continuous and/or categorical independents; to determine the effect size of the independent variables on the dependent; to rank the relative importance of independents; to assess interaction effects; and to understand the impact of covariate control variables. The impact of predictor variables is usually explained in terms of odds ratios.

Note that logistic regression calculates changes in the log odds of the dependent, not changes in the dependent itself.

Binary logistic regression predicts the "one" value of the dependent, using the "zero" level as the reference value.

Table 2 Odds ratio for b coefficients in the model and Wald statistic significance value

	OBJET9MOD	OBJET10MOD	OBJET11MOD	OBJET13MOD
SIZE	1.662 (0.054)	1.5 (0.068)	8.551 (0.027*)	1.244 (0.183)
F. R&D	1.629 (0.208)	0.976 (0.873)	2.919 (0.043*)	1.239 (0.41)
EXPORT	1.13 (0.43)	1.201 (0.245)	1.407 (0.038*)	1.505 (0.014*)
Intercept	1.273 (0.095)	1.041 (0.775)	1.634 (0.036*)	1.136 (0.364)

* Significant if $p < 0.05$

Intercepts are the log odds of the dependent when predictors are at their average values.

Wald statistic significance and the odds ratio are shown in Table 2. Odds ratios are effect size measures. The odds ratio is the natural log base, e, to the exponent of the parameter estimate, b, and, for continuous variables, the odds ratio represents the factor by which the odds (event) change for a one-unit change in the variable.

Results show that independent variables are significant only the model found for OBJET11MOD in determining whether a company is high or medium oriented toward environmental impact while innovating or rather have low or not orientation.

From the model we can found that SIZE has the greater impact in changing the odds of being high/medium oriented rather than low/not oriented. We may say that when the independent variable increases one unit, that is the company has a value in the variable SIZE obtained one standard deviation higher than the mean value for the sample, the odds that the dependent is equal to 1, increase by a factor of 8,55, when other variables are controlled.

Then we can conclude that, size, formal R&D Activities and Export orientation influence with different impact on the environmental orientation of the firms in the automobile industry while innovating.

Variables studied show that are not able to discriminate between companies in the automobile industry that set the reduction of materials and energy per unit as an objective of high/medium importance or low/not importance while innovating. Neither for the objective of accomplishing environmental legislation requirements.

5 Conclusions

In this study have been emphasized a wide range of actions that a leader automotive company like Faurecia, has been taken over the past few years. This chapter reflects the information collected after the interview with eco-design experts from Faurecia Group to detect environmental objectives followed while the innovating process is taking place, mainly in the design phase.

Observed results are reinforced by the empirical results that show that environmental orientation is influenced by the company's characteristics.

For the Spanish automotive firms, the study has detected that environmental proactivity while innovating is determined mainly by the size of the firms, measured by the total income, total investment, size R&D investment and R&D employees, and also, but less, by the formal R&D activity (number of patents) and export orientation.

Accordingly with the results, bigger companies with higher number of patents and with a wider international presence are more likely to be environmentally oriented when they are innovating. As automotive firm's innovations are focused and take part mainly on the design phase of products, we can conclude that eco-design is more likely to take part in big companies with high external and innovation orientation and that these companies are eco-innovation drivers throughout the automotive industry.

The study also has found no significant differences on companies' characteristics attending the importance of other aspects like energy and material reduction or environmental legislation accomplishment while innovating. Although, energy and material reduction might be related to environmental innovation, they are also highly influenced by operational facts, so company orientation might be affected by other variables like economic performance, costs structure or its financial situation.

Further research will continue to explore the specific characteristics of the automotive industry when facing eco design and other related environmental issues.

Acknowledgments The authors would like to thank the Universitat Politècnica de València for supporting the UPV-PAID 2011-1879_ First Projects Funding and the sabbatical research of M. Segarra. Also the Spanish Economy and Competitiveness Ministry for its financial support through the research project (EC02011-27369), and Faurecia for providing helpful information.

References

Acs ZJ, Anselin L, Varga A (2002) Patents and innovation counts as measures of regional production of new knowledge. Res Policy 31(7):1069–1085

Bohdanowicz P (2005) European hoteliers' environmental attitudes: greening the business. Cornell Hospitality Quart 46(2):188–204

Becker MC, Lazaric N, Nelson RR, Winter SG (2005) Applying organizational routines in understanding organizational change. Ind and Corp Change 14(5):775–791

Churchill NC, Lewis VL (1987) The five stages of small firm growth. Long Range Plan 20(3):45–52

Clark G, Kosoris J, Hong LN, Crul M (2009) Design for sustainability: current trends in sustainable product design and development. Sustainability 1(3):409–424

Coulter S, Bras B, Winslow G, Yester S (1998) Designing for material separation: lessons from automotive recycling. Trans Am Soc Mech Eng J Mech Des 120:501–509

Criado E (2007) Reflexiones sobre el futuro de la Industria Europea de la Cerámica. Boletín de la Sociedad Española de Cerámica y Vidrio 46(1):39–44

Crosby M (2007) Patents, innovation and growth. Econ Rec 76(234):255–262

De Medina HV (2006) Eco-design for materials selection in automobile industry, In: Proceeding of the comunicação técnica elaborada para o 13th CIRP internacional conference on life cycle engineering, LCE 2006—towards a closed loop economy, Leuven, pp 299–304

Dubé L, Paré G (2003) Rigor in information systems positivist case research: current practices, trends, and recommendations. Mis Quart 27(4):597–636

Encaoua D, Guellec D, Martínez C (2006) Patent systems for encouraging innovation: lessons from economic analysis. Res Policy 35(9):1423–1440

European Commission (2000) Environmental data centre on Waste_Data Directive 2000/53/EC, epp.eurostat.ec.europa.eu› European Commission

Eurostat (2011) End-of-life vehicles: Detailed data, http://appsso.eurostat.ec.europa.eu/nui/show.do

Ferrao P, Amaral J (2006) Assessing the economics of auto recycling activities in relation to European union directive on end of life vehicles. Technol Forecast Soc Change 73(3):277–289

Gerrard J, Kandlikar M (2007) Is European end-of-life vehicle legislation living up to expectations? Assessing the impact of the ELV Directive on "green" innovation and vehicle recovery. J Clean Prod 15(1):17–27

Greiner LE (1997) Evolution and revolution as organizations grow: a company's past has clues for management that are critical to future success. Family Bus Rev 10(4):397–409

Hair JF, Black WC, Babin BJ, Anderson RE, Tatham RL (1998) Multivariate data analysis. Prentice Hall, New Jersey

Jorgenson AK, Burns TJ (2007) The political-economic causes of change in the ecological footprints of nations, 1991–2001: a quantitative investigation. Soc Sci Res 36(2):834–853

Leduc G, Mongelli I, Uihlein A, Nemry F (2010) How can our cars become less polluting? An assessment of the environmental improvement potential of cars. Transp Policy 17(6):409–419

Manzini E, Vezzoli C (1998) Agenzia nazionale per la protezione dell'ambiente. Lo sviluppo di prodotti sostenibili: i requisiti ambientali dei prodotti industriali. Maggioli, Italy

Miret Pastor L, Segarra Oña M, Peiró Signes A (2011) Cómo medimos la ecoinnovación?: análisis de indicadores en el sector turístico (how to rate ecoinnovation?: A tourism sector indicator's analysis), Tec Empresarial 5(2):15–25

Munksgaard J, Wier M, Lenzen M, Dey C (2005) Using input_output analysis to measure the environmental pressure of consumption at different spatial levels. J Ind Ecol 9(1–2):169–185

Nelson RR, Winter SG (2002) Evolutionary theorizing in economics. J Econ Persp 16:23–46

Orsato R, Wells P (2007) The automobile industry and sustainability. J Cleaner Prod 15(11–12):989–993

Salomon R (2005) Learning from exporting: new insights, new perspectives. Edward Elgar Publishing, Cheltenham

Santini A, Morselli L, Passarini F, Vassura I, Di Carlo S, Bonino F (2010) End-of-life vehicles management: Italian material and energy recovery efficiency. Waste Manage (Oxford) 31(3):489–494

Schiavone F, Pierini M, Eckert V (2008) Strategy-based approach to eco-design: application to an automotive component. Int J Veh Des 46(2):156–171

Segarra-Oña MV, De-Miguel-Molina M, Payá-Martínez A (2011a) A review of the literature on eco-design in manufacturing industry: Are the institutions focusing on the key aspects? Rev Bus Inform Syst 15(5):61–67

Segarra-Oña M, Peiró-Signes A, Albors-Garrigós J, Miret-Pastor P (2011b) Impact of innovative practices in environmentally focused firms: moderating factors. Inter J Environ Res 5(2):425–434

Wackernagel M, Onisto L, Bello P, Callejas Linares A, López Falfán I, Méndez García J, Suárez Guerrero A, Suárez Guerrero G (1999) National natural capital accounting with the ecological footprint concept. Ecol Econ 29:375–390

Wackernagel M (2002) T Tracking the ecological overshoot of the human economy. Proc Natl Acad Sci USA 99(14):9266–9271

Whitmarsh L, Köhler J (2010) Climate change and cars in the EU: the roles of auto firms, consumers, and policy in responding to global environmental change. Cambridge J Reg Econ Soc 3(3):427

Wimmer W, Lee KM, Quella F, Polak J (2010) Outlook: sustainability–what does the future hold? ECODESIGN–Compet Adv 18:191–194

Sustainability Issues for Vehicles and Fleet Vehicles Using Hybrid and Assistive Technologies

Lindita Prendi, Simon Che Wen Tseng and Edwin K. L. Tam

Abstract Hybrid electric vehicles (HEVs) are considered preferred alternatives to internal combustion engine vehicles because they can reduce air emissions and fuel consumption while performing competitively against conventional vehicles in commuter usage scenarios. Hybrid vehicle technologies vary widely and offer different advantages and disadvantages from environmental and socio-economic perspectives for passenger vehicles. At the other extreme, fleet vehicles are operated differently from passenger vehicles and idle for about 70 % of their operation time. Hybrid vehicles have yet to be utilized widely by fleets: they would appear to complement fleet operations but there are other approaches to reduce emissions, including assistive technologies to operate in-vehicle equipment and maintain fleet vehicle capabilities instead of idling. Hybrid vehicles and assistive technologies, such as auxiliary power units could offer significant benefits to fleet vehicles by powering electronics while idling and thus reduce the need for conventional engine operation. However, do hybrids and assistive technologies actually provide justifiable benefits in passenger vehicles and fleet vehicles? There are specific end-of-life issues with hybrids and assistive technologies that should be assessed. These issues and the overall sustainability of vehicles can be assessed using life cycle assessment (LCA) approaches.

Keywords Hybrid · Electric · Batteries · Life cycle assessment · Emissions · Fuels · Diesel · Idling · Fleets · Vehicles

L. Prendi (✉) · E. K. L. Tam
Department of Civil and Environmental Engineering,
University of Windsor, Windsor, ON, Canada
e-mail: edwintam@uwindsor.ca

S. C. W. Tseng
Department of Mechanical, Automotive and Materials Engineering,
University of Windsor, Windsor, ON, Canada

P. Golinska (ed.), *Environmental Issues in Automotive Industry*,
EcoProduction, DOI: 10.1007/978-3-642-23837-6_4,

1 Introduction

Mobile sources (cars and trucks) are major contributors to air pollution in metropolitan cities (Calef and Goble 2007). According to Environment Canada (2010) the transportation sector is the largest contributor of GHG emissions in Canada. It has also been reported that transportation sector contributes for one-fourth of GHGs and air pollutants on a global scale (Ogden et al. 2004). With the increase of population the number of vehicles on the road has correspondingly increased. Air pollution from vehicle exhaust is also known to contribute to respiratory system illnesses. Hybrid vehicles are considered a preferred alternative to internal combustion engine vehicles because they are reported to reduce air emissions and fuel consumption while remaining competitive in performance with conventional vehicles in urban or commuter usage scenarios. It has been estimated that "hybrid vehicle sales in the USA exceed 1 million per year by 2012" (ACA 2008). There are a number of variations on hybrid technologies however, ranging from different fuel formulations to different powertrain combinations. The different hybrid configurations can influence how they in turn affect environmental, social, and economic considerations.

At the other extreme of vehicle use, environmental concerns about emissions and rising fuel costs are driving fleet operators (i.e. police, ambulance) to consider alternative technologies for their fleets. Extended fuel consumption and air emissions are attributed to the unique operations of fleet vehicles and in particular, during idling. While drivers of passenger vehicles may have the option of simply not idling, fleet operations, and in particular, emergency vehicle operators, may need to keep the vehicle operating to provide power to operate critical onboard equipment. These demands may be exacerbated during seasonal, temperature extremes. However, prolonged idling can impose significant environmental and economic burdens. Hybrid vehicles might be an attractive solution, but have yet to be utilized widely by fleets. There are other, increasingly mature approaches to reduce emissions, including idling reduction or assistive technologies to operate in-vehicle equipment and maintain fleet vehicle capabilities instead of idling.

This chapter summarizes the state-of-the-art hybrid and assistive technologies for vehicles in general and fleet vehicles, and presents the environmental and socio-economical issues associated with these innovative hybrid and assistive technologies. Substantial research has been conducted on the performance of vehicles using different power trains and energy sources. There is significantly less available research on assistive technologies and life cycle environmental trade-offs for fleet vehicles, especially when considering their unique modes of operation and the novel technologies required for operational purposes. The chapter will focus on the following major aspects:

1. The issues, benefits, and impacts from various hybrid vehicle technologies, including variations in terms of fuels and powertrains;
2. Socio-economic effects related to hybrid vehicles; and

3. Idle reduction or assistive technologies and fleet vehicle issues, including end-of-life issues that may have to be considered.

2 Overview of Hybrid Technologies

The term "hybrid vehicle" is popularly thought of as a combination of a gasoline internal combustion engine and an electric motor. However, there are a variety of technologies that can qualify a vehicle as a "hybrid". "Hybrid as it affects vehicles could be in terms of the fuel used in the internal combustion engine of vehicles (fuel hybridization) or the combination of propulsion power from an internal combustion engine with that produced by electric energy stored in batteries (drivetrain hybridization)" (Momoh and Omoigui 2009). Since its inception, automobiles have been powered by either the internal combustion engine (ICE) or by an electric motor. Electric motors offered an unparalleled quietness in operation and zero tail pipe emissions, but previously lacked the technology to provide the performance demanded by users.

ICEs were simpler in design and with the abundance of inexpensive crude oil at the time, secured its dominance until recently. With the increase in global awareness in the environmental impacts of personal transportation and the rising cost in crude oil, the trend is to move away from ICE and towards electric vehicles (EV). However, the electric vehicle has not yet become a dominant choice partly due to the still developing battery technology. The hybrid electric vehicle (HEV) is now seen as the intermediate step towards the eventual goal of having EVs serve as the primary type of passenger vehicle (Katrasnik 2007).

Fuel hybridization utilizes current fueling infrastructures by using various different fuels and fuel mixtures to enhance ICE combustion. This outcome will increase thermal efficiency and reduce tail pipe emissions and fuel consumption. Drivetrain hybridization incorporates both the ICE and the electric motor (EM) in several configurations to promote steady state engine operation by minimizing transient operation/combustion variations. The increase in mechanical efficiency of the ICE results in lower tail pipe emissions and fuel consumption (Stone 1999). The following section reviews the state-of-the-art technology in both fuel and powertrain hybridization.

2.1 Fuel Hybridization

Fuel hybridization can be performed in several ways:

1. Using alternative fuels in current 4-stroke ICE in two forms:
   a. gaseous—hydrogen, methane (LNG, CNG)
   b. liquid—methanol, ethanol, bio-diesel

2. Using fuel blends that mix different alternative fuels with fossil fuels (e.g., E10).

The alternative fuels reviewed include: hydrogen, propane, methane, methanol, ethanol and bio-diesel. All these fuels are produced from renewable sources to lower the dependence on the traditional fossil fuels. Methane and propane are produced by anaerobic reactions during waste disposal/management systems. (i.e., landfills). Methanol and ethanol are produced by fermentation of starchy crops such as corn. Bio-diesel is produced by chemically reacting lipids with an alcohol, biomass to fuel or most recently algae to fuel reactions (Demirbas and Demirbas 2011). There have been concerns however, that the use of food crops to produce fuel could result in food shortage. As a result, recent developments have moved toward alternative fuel production using non-food crops such as grass, wood, and algae (Kamimura and Sauer 2008).

Replacing fossil fuels with alternative fuels such as methane or ethanol, is not ideal for conventional spark ignition (SI) engines (Pourkhesalian et al. 2010). It has been reported that volumetric and thermal efficiencies are the highest in gasoline fueled ICEs while brake specific fuel consumption (BSFC) is lowest with gasoline. However, there are some benefits in reduced emissions when using alternative fuels instead of gasoline (Pourkhesalian et al. 2010). Currently spark ignition ICE are designed to operate with gasoline, but if a fuel specific ICE design is developed then alternative fuels might be more efficient. The limiting factor for alternative fuel replacement is that current production capabilities cannot mass produce sufficient quantities of alternative fuels to replace gasoline (Pourkhesalian et al. 2010).

For conventional diesel/compression ignition (CI) engines, replacing fossil fuels with alternative fuels (bio-diesel) is more promising than SI engines (Pourkhesalian et al. 2010). For most current CI engines, minor or no modifications are necessary to operate them with bio-diesel. Bio-diesel produces lower emissions of NOx, PM, and CO and come from renewable recourses. However, BSFC is higher compared to regular diesel due to a lower heating value (Lapuerta et al. 2008). Existing infrastructures can deliver and dispense fuel to vehicles with minor to no modification. Similar to SI alternative fuels, the challenge with bio-diesels is that the current production capabilities cannot mass produce sufficient bio-diesel to replace regular diesel supplies.

Gasoline fuel blends are divided into flex fuel engine blends or ethanol blends. Some of the vehicles currently available can operate with gasoline and other alternative fuels such as natural gas (primarily methane). Compressed natural gas (CNG) is usually used to supplement gasoline in fleet vehicles that idle for a long period of time to reduce the tail pipe emissions. This reduction is due to the smaller hydro-carbon (HC) chains with methane compared to gasoline, which is easier for complete combustion. Current pump gasoline has between 10 and 15 % ethanol blended in to increase the octane number (ON) while reducing refinery processes by utilizing the higher ON of methane and its latent heat of evaporation. However, the overall heating value is lower than pure gasoline, which will increase

BSFC. Both fuel blend methods are aimed to promote more efficient combustion to reduce tail pipe emissions while lower the dependence on fossil fuels (Delgado et al. 2007).

Diesel blends are derived by mixing bio-diesel with regular diesel. Overall, the combustion of bio-diesel in CI engines produces lower smoke, PM, CO, and HC compared to regular diesel fuel with the same (if not improved) engine efficiencies. However, with the increase in blend percentage, there is an increase risk of lower durability due to moving parts sticking, injector choking and filters blocking (Fazal et al. 2011). Blends can range from B10 to B20 in current pump diesel up to B80 in experimental engines. Similarly, bio-diesel reduces the tail pipe emission by promoting complete combustion with its inherent oxygen molecules in the fuel. This lowers pyrolysis of the fuel and lowers PM and CO creation. However, there is a slight increase in NOx due to the excess oxygen. Catalytic technologies such as a lean NOx trap (LNT) can capture the excess NOx produced (Stone 1999).

2.2 Drivetrain Hybridization

Drivetrain hybridization utilizes a combined propulsion system from both:

- Chemical (fuel) energy release from an ICE; and
- Electrical energy generated by the ICE mechanical system/EM stored in batteries.

This combination benefits from both propulsion systems' advantages. From the ICE standpoint, the advantages include the superior power density of carbon-based fuel, the simplicity and low cost of design and manufacture, and the ability to use existing infrastructure. From the EM standpoint, the advantages include instantaneous torque and waste energy recovery (regenerative braking). Apart from utilizing advantages of individual systems, the two systems supplement each other and work together synergistically. The ICE on its own suffers from inefficiencies during transient operations, for example during stop and go traffic conditions with frequent acceleration and deceleration. By incorporating an EM into the propulsion system, engine rpm fluctuations can be minimized during acceleration to promote more steady state ICE operations that increase the overall mechanical and volumetric efficiencies. Regenerative braking technology can be implemented to capture the conventionally wasted mechanical energy during braking/deceleration by charging a traction battery on board to be used for acceleration assistance. The instantaneous torque provided by the EM reduces the necessary ICE engine size and can compensate for the lack of torque at the lower rpm of the ICE engine. This further increases the ICE's mechanical efficiency by reducing the parasitic losses inherent to it (for example, due to less rotating mass and mechanical friction). The further implementation of technologies such as integrated starter generator (ISG) and cylinder deactivation enhances the overall propulsion system's efficiency by

Table 1 Types of drivetrain hybridization and typical applications

	Series hybrid	Parallel hybrid	Series–Parallel hybrid
Make	Chevrolet (GM)	Honda	Toyota
System	Two mode hybrid system	Integrated motor assist	Hybrid synergy drive
Model	Volt	Insight/Civic	Prius

minimizing wasted energy inherent to the conventional ICE system (Momoh and Omoigui 2009).

There are many ways to configure drivetrain hybridization. The specific roles of both ICE and EM within the propulsion system determine the type of hybridization implemented and can be categorized into three main groups: (1) series hybrid; (2) parallel hybrid; and (3) series–parallel hybrid (Table 1). Each design has its respective strengths and weaknesses.

2.2.1 Series Hybrid

The role of an ICE engine is not to power the drivetrain directly via conventional crankshaft rotations. Instead, the electric motor uses the electrical energy generated by ICE to power the drivetrain propelling the vehicle. There is no mechanical connection between the ICE and the drivetrain. The instantaneous torque provided by the EM is more efficient than parallel hybrid systems in stop and go traffic conditions. However, this configuration requires separate motor and generator portions, which usually has a lower combined efficiency compared to conventional transmissions that offsets the overall vehicular efficiency (Momoh and Omoigui 2009).

The state-of-the-art of series hybridization in production is the Chevy Volt. The system, known as Range Extender system, uses the ICE solely to charge the traction battery pack and vehicle propulsion is done by the EM. Current research and development (RandD) efforts focus on applications of 'in-wheel' motors where an EM is installed at each wheel driven by the battery. This configuration enables variable wheel speeds that essentially provide the functionality of both the all-wheel-drive (AWD) and limited slip differential (LSD) systems without the conventional mechanical systems (Rambaldi et al. 2011).

2.2.2 Parallel Hybrid

Both the ICE and EM are connected to the drivetrain with the EM functioning in a supplementary role. The ICE is the main source of propulsion with assistance from the EM only under heavy loads such as in the case of acceleration. The EM is usually positioned between the ICE and the conventional transmission. According to Schouten et al. (2003) there are five ways to operate the system depending on the power flow desired:

1. Provide power to the wheels with only the ICE;
2. Use only the EM;
3. Use both the ICE and the EM simultaneously;
4. Charge the batteries using part of the ICE power to drive the EM as a generator (the other part of ICE power is used to drive the wheels); and
5. Slow down the vehicle by letting the wheels drive the EM as a generator that provides power to the battery, otherwise known as regenerative braking.

The state-of-the-art of parallel hybridization in production is Honda's Integrated Motor Assist (IMA) system. Depending on driving conditions and driver inputs, the EM can assist the ICE under heavy load as in acceleration, charge the battery while under light loads as in cruising, or charge the battery by regenerative braking. Current research and development efforts are focused on designing better energy management controllers to optimize the operational efficiencies of all the components. These components include the ICE, EM, battery state-of-charge (SoC), EM/generator speed, braking and gear shifting (Schouten et al. 2003).

2.2.3 Series–Parallel Hybrid

This configuration combines series and parallel hybrid systems: both motors can power the drivetrain independently. The combined power output is controlled by a power splitter where 0–100 % of power from either motor can be utilized in any ratio depending on driving conditions (e.g., 50 % ICE and 50 % EM). This system is also known as "power split" hybridization (Schouten et al. 2003).

The state-of-the-art of series–parallel hybridization in production is the Toyota Prius Synergy Drive Hybrid System. Series hybrid characteristics are used at engine start and low speed acceleration and parallel hybrid characteristics are used during high speed acceleration and braking. This system has the highest overall efficiency. However, there are extra components, complexity, and costs associated with it. Current research and development efforts are focused on component downsizing (e.g., the power convertor without compromising energy density) power loss/heat management (e.g., more heat resistant modules and simplified cooling) and finally cost reduction through component standardization (Mastumoto 2005).

2.2.4 Internal Permanent Magnet Motors

Internal Permanent Magnet (IPM) motors, and especially double intelligent power module (IPM) motors, are currently the state-of-art traction motors in HEV applications. Numerous requirements are crucial for their successful operation. Some of these include: high torque and power density, high starting torque, high power at cruising speed, short term overload capacity, low acoustic noise, low torque ripples, maximum variation of d-q axis inductance, least magnet flux leakage, temperature and surface corrosion constraints, excessive open circuit back-emf, and load and no-load stator iron loss at high speeds (Rahman 2008). Apart from control modules and inverters, the V style IPM design utilized by Toyota has been proven superior to both the induction and reluctance motors in terms of electric torque characteristics (Rahman 2008).

2.3 Battery Technologies for Hybrid Vehicles

The earliest automobiles were operated mechanically without the use of electronics. Engines were manually started by cranking the engine with a handle attached to the crankshaft, which in turn rotated the spark plug cap and rotor to initiate engine operation. Since then, automotive technologies have advanced significantly and become more sophisticated, particularly with the addition of electronics. These advancements led to automobiles that produce more power, use less fuel, and are user friendly. Automotive batteries became a necessity on vehicles to start the engine and to power onboard electronics when the engine is not running.

The efficiency of hybrid electric vehicles depends heavily on the capacity of the batteries equipped. Batteries used in electric vehicles are also referred to as "traction batteries". This efficiency impacts directly the fuel economy of the HEV. As such, battery technology has been an essential research and development topic since the conception of both electric and hybrid electric vehicles. Current battery systems under development are: Nickel Metal Hydrate (NiMH), Lithium ion (Li ion), Lithium Metal Polymer (Li MP), Zebra (Sodium Metal Chloride), and Nickel Zinc (NiZn). Each system has advantages to its design (Wehrey 2004).

2.3.1 NiMH: Nickel Metal Hydrate

Currently, NiMH is the most widely used system in both HV and HEV productions; it is the benchmark for automotive battery systems. It has been successfully implemented with hundreds of thousands of miles logged on both test and production vehicles. Its reliability and relatively low cost are the reason for its tremendous success. However, the weight of this battery type is heavier compared to other designs, which adds to the overall weight of the vehicle (Wehrey 2004).

2.3.2 Li-ion: Lithium Ion

Li-ion batteries will be utilized for the first time on a mass produced automobile in the 2011 Chevy Volt and the 2012 Honda Civic Hybrid (Honda World Wide 2011). The design has superior characteristics in terms of power, energy density, size, weight and performance. With maturing technology, the original high cost is decreasing. Furthermore, the ability to make traction batteries from smaller cells further encourages automotive applications.

2.3.3 Other Battery Types

There are other emerging battery technologies. The Lithium Metal Polymer (Li MP) battery design is not aimed at the HEV industry, but rather EV, telecommunication, or stationery equipment. However, successful EV prototypes have been made such as the Think City EV.

The Zebra (sodium metal chloride) is a high temperature battery targeted for EV applications and can withstand freezing without adverse effects on its cycle. It also has four times the specific energy of conventional lead-acid batteries at 120 Wh/kg. Furthermore, it has no shelf-life issues common to all other battery designs (Wehrey 2004).

Nickel Zinc (NiZn) batteries are aimed at secondary battery applications. This design has not been received well due to its tendency to form dendrites at the zinc electrode. Further developments are needed before this technology can find wide applications.

2.4 Battery Related Performance

Many of the concerns with hybrid vehicles relate to battery performance. Liaw and Dubarry (2007) studied the driving cycle and battery performance of hybrid vehicles in real-life scenarios. The authors suggest that battery performance tests conducted in the laboratory do not always represent what happens during real-life operations. Energy consumption depends on ambient operating conditions and these conditions are difficult to control and therefore measure. The driving cycle term used from this group of authors refers to the speed versus time relationship, while the duty cycle term refers to the power versus time relationship. The road conditions and driving behavior were accounted for and described by the driving pattern term. The challenge is to then correlate the theoretical battery performance with HEV usage in real-life situations.

Liaw and Dubarry (2007) used fuzzy logic pattern recognition (FL-PR) for their analysis of data collected by 15 Hyundai Santa Fe battery-powered electric sport utility vehicles. During the urban driving cycle (stop and go), the traffic and road conditions had an increasing impact on the effectiveness of energy use. The effective force (EF = kWh/km) measured the vehicle performance increased under these conditions. However, it was constant under rural and highway driving scenarios. The authors report that the driving event does not have to match the road type. For example, the stop and go scenario is relevant for urban roads but a driver might encounter the same pattern on a busy highway during rush hour.

The intensity of the peak power and frequency of occurrence are two important factors that affect battery performance and life. Average power and energy consumption can also be used to access the performance and life of batteries. Driving distance and energy consumption showed a linear relationship. An increase in energy consumption with driving scenarios (urban, highway etc.) was not monotonic. In terms of peak power, there is a monotonic increase with the driving scenario. As a result, there is an increase of the peak power with driving scenarios going from local to highway settings (Liaw and Dubarry 2007).

The batteries or any other electrical energy storage unit are sized to achieve appropriate energy (kWh) and peak power (kW) so that the vehicle meets the required performance (Burke 2007). The battery cycle life for deep discharge is an important factor since the batteries are regularly deep discharged. This is true for the electric vehicles and the battery is sized based on the range of vehicle travel. However, for hybrid-electric vehicles, the batteries are sized based on the peak power from the unit during acceleration. The battery stores an amount of energy that is larger than what is needed, but this additional energy allows the battery "to operate over a relatively narrow state of charge range (often 5–10 % at most)". As a result, batteries used in HEVs have longer life and longer cycle life (Burke 2007).

Power density, which is the maximum power the battery can supply, differs from energy density, which is how long the battery can supply the power. There is a tradeoff between these two parameters. In vehicle designs where the battery is used for driving and the engine is used only for high power demands or traveling at speeds above the normal specifications, the fuel economy will be improved. In addition the energy and power requirement will be lower and as a result batteries can be less expensive and smaller in size. Burke (2007) reports that nickel metal hydride (NMH) batteries are the most common types used in hybrid vehicles.

Batteries designed for HEVs are smaller in cell size, have higher power capability and smaller weight because the transfer of energy in and out of the battery should be very efficient. However, the high power density is achieved by trading off the energy density values which in HEV batteries are lower than other vehicles (i.e. EVs). Burke (2007) compared the ability of batteries to withstand charge/discharge cycles between conventional batteries and HEV batteries. In these vehicles the battery was only used to make the engine more efficient and recover energy during braking. There was a 50 % fuel economy improvement compared to conventional gasoline engines. The main engine can be down-sized in "full" hybrid vehicles where the electric motor is larger (50 kW or larger). Nickel metal hydride or lithium-ion batteries can be used and are sized by the power demand. In these vehicles the battery is shallow discharged at an intermediate state of charge (Burke 2007).

3 Hybrid Vehicles: The Environment and Society

The need to reduce the dependency on petroleum based fuels and air pollution concerns has been the driving force toward clean vehicle technologies. In California it has been reported that 51 % of NO_x emissions is attributed to "on-road mobile sources" (Calef and Goble 2007). Several studies have reviewed the potential negative health effects of air vehicle exhaust from emissions such as SO_2, NO_x, and O_3. A number of alternative technologies have the potential to address environmental, economical and social issues associated with ICEVs (internal combustion engine vehicles) such as hybrid-electric vehicles (HEVs), fuel cell vehicles (FCVs) and battery electric vehicles (BEVs) (Turton and Moura 2008). Another potential benefit of HEVs is the control of the place and time that emissions occur: although they are not designed to operate completely using an electric motor, they can be driven as zero emission vehicles if the right battery type is chosen (Calef and Goble 2007).

3.1 Fuel Consumption and Emission Issues

In the late 1990s there was substantial evidence that hybrid vehicles provided improvements in fuel efficiency and reduced carbon dioxide and nitrogen oxide emissions (O'Dell 2000; Easterbrook 2000). Duoba et al. (2005) reported that hybrids were the most fuel efficient vehicles on the market at the time. As an example, the fuel economy of Honda Civic hybrid was rated as 42 mpg (city and highway cycle) while the conventional Honda Civic was 25 mpg (EPA 2008).

D'Agosto and Ribeiro (2004) report that hybrid buses would decrease fuel consumption by more than 20 % and as a result, reduce fuel costs and air emissions. They examined the cost of converting the existing bus fleet with hybrid versions and reported that fuel savings offset the initial increased investment. It was reported that fuel savings between 35 and 40 % were more likely to occur at speeds between 10 and 15 km/h (D'Agosto and Ribeiro 2004). These speeds are very frequently the common travel speeds during traffic jams in metropolitan cities. Furthermore, the highest cost related to conversion of conventional buses to hybrid ones is the battery cost.

Gasoline, hybrid-electric and hydrogen-fueled vehicles were compared in terms of greenhouse gas emissions by Uhrig (2006). CO_2 was used as the main greenhouse gas as the author suggests that it has higher impact and longer residence time compared to other pollutants (i.e. CH_4, NO, CO etc.). It was found that CO_2 emissions for gasoline ICE vehicle are directly related to fuel use. The results show that there was an increase in fuel mileage and decrease in CO_2 emissions respectively. However, these HEVs still depend on fossil fuels and so there is need to develop new technologies in order to reduce dependency on fossil fuels (Uhrig 2006).

Mizsey and Newson (2001) compared several power trains for well-to-wheel efficiencies, CO_2 emissions, and the investment costs. The gasoline internal combustion engine (ICE) was used as a baseline for comparison and the other alternatives considered were hybrid diesel, fuel cell operating with hydrogen produced on a petrochemical basis, methanol reformer-fuel cell system, and gasoline reformer-fuel cell system. The gasoline ICE had the highest CO_2 emissions, but the cost for the gasoline engine powertrain is the lowest. If the vehicles are therefore compared only on environmental basis the ICE gasoline powertrain has the worst performance. In terms of the well-to-wheel efficiency the hybrid diesel comes out first: it has the lowest CO_2 emissions along with the compressed hydrogen produced by natural gas technology (Mizsey and Newson 2001). These findings are encouraging for developing hybrid diesel vehicles and their commercialization, which to date, has not enjoyed the same visibility as hybrid gasoline-electric vehicles.

Vehicle exhaust contains a number of greenhouse gases and contaminants identified as contributors to air pollution. Significant advancements have been made through after-treatment technologies in order to reduce the effect of vehicle exhaust to the environment. Alternative technologies and fuels have been investigated and implemented in order to achieve emission reduction while maintaining vehicle performance. It has been reported that through improvements in technology, it has been possible to reduce emissions up to 95 %. Karman (2006) emphasizes that these alternative technologies and fuels should be carefully evaluated through life-cycle analysis (LCA) in order to claim their real benefits.

3.2 *Using Life Cycle Assessment to Evaluate Hybrid Technologies*

Life cycle assessment or analysis (LCA) has emerged as key tool for comparing the performance of different products or processes, both before/after changes and as compared to other products/processes, because of its ability to assess a much broader context. The LCA framework is illustrated in Fig. 1. As an example of the benefits of using LCA for evaluating hybrid vehicle technologies, Karman (2006) used GHGenius, GREET, and CSIRO LCA models to compare greenhouse gas emissions between diesel and compressed natural gas (CNG) buses in Beijing. Even though natural gas yielded lower CO_2 emissions during the operation stage and also from upstream operations compared to diesel, natural gas buses produced more CH_4. CH_4 has a 21:1 greenhouse gas potential ratio (CH_4:CO_2) when compared to CO_2. A complete LCA thus revealed that the total CO_2-equivalent life-cycle effect was higher for CNG than for diesel (Karman 2006).

The benefits of LCA for evaluating environmental impacts of transportation systems have also been discussed by Stanciulescu and Fleming (2006). The authors also give a detailed description of the GHGenius model which was

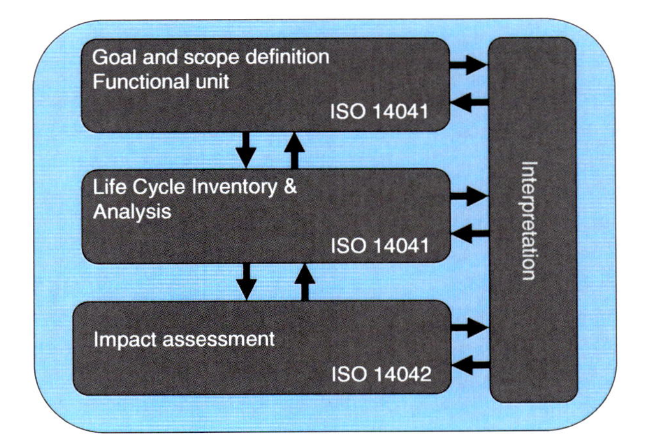

Fig. 1 LCA framework (adapted from ISO 14040 2006)

developed for Natural Resources Canada and contains detailed information related to fuel cycles and can be used to model and predict the environmental impact of conventional and alternative fuels and propulsion systems. The model has been successfully used by the government and industries in many studies, and is best fitted for Canadian scenarios. The GHGenius is capable of analyzing emissions from different vehicle and fuel combinations, which makes it an attractive model for researchers.

3.3 Diesel Benefits and Impacts

Diesel vehicles have not yet attracted widespread consumer attention in North America (Albert et al. 2004). Some of the reasons are related to the soot and noise generation and NO_x emissions. However, advancements in technology have targeted some of these issues and as a result there have been many improvements. It has been reported that diesel engines are more efficient than gasoline engines and the use of turbo charging can increase the performance of diesel engines so that they are comparable to gasoline engines.

The fuel pathways for gasoline, diesel, biodiesel and ethanol blended diesel are particularly critical. Interestingly, Stanciulescu and Fleming (2006) showed that the energy use for well-to-tank case was the same for both diesel and biodiesel fuel production stage. The same trend was also observed about the GHG emissions from these two fuels during the well-to-tank stage. However, both diesel fuels show lower energy use and GHG emissions than reformulated gasoline with 30 ppm sulfur content. Most of GHG emissions are released during the vehicle operation stage from reformed gasoline ICE and hybrid vehicles: the same can be said for diesel and diesel hybrid types. However, in both gasoline and diesel cases the hybrid alternative has lower GHG emissions during the operation stage. Overall out of these four scenarios it is the diesel hybrid that yields the lower GHG emissions in g/km CO_2 equivalents (Stanciulescu and Fleming 2006).

The majority of hybrid electric vehicles (HEVs) produced to date have a gasoline engine. In the last few years however diesel hybrids have been the focus of research (Albert et al. 2004). One benefit of these hybrid vehicles is the high flashpoint of diesel which adds to the safety of the vehicle in a collision scenario. The authors used ADVISOR to simulate the fuel efficiency of large and small diesel hybrid vehicles. It was reported that for the large hybrid SUV there was a 25 % improvement in fuel economy for a hybrid factor of 0.1. The hybrid factor (HF) is the ratio of the power of electric motor over the power of both electric motor and main engine combined. The economy increased with the increase in HF, but the performance suffered. The optimum fuel economy (21.6 mi/gal) was achieved at HF = 0.6 and indicated a 44 % improvement compared to the conventional SUV. However, the maximum speed that could be achieved was 82 mph. For the small SUV an HF of 0.6 was still the optimum with a fuel economy of 26.1 mi/gal and a 97 % increase from conventional vehicle; however the maximum speed was 81.6 mph (Albert et al. 2004).

The output power of the battery was compared with the output power of the electric propulsion motor. It was found that fuel economy was not improved when the output power of the battery units was matched to the power of the electric motor. Conversely, the fuel economy decreased due to the weight increase caused by addition of battery units. The maximum speed did not increase significantly, but this is in contrast to the fuel economy which decreased. The performance factor that was improved significantly was the gradeability, or the ability of the vehicle to ascend a slope, because of the addition of the batteries. The authors suggest that for hybrid diesel engine vehicles, the tradeoff between increased graedability, maximum speed and acceleration against the fuel economy make hybrid diesel engine vehicles as an attractive choice (Albert et al. 2004).

Turton and Moura (2008) also examined the potential of using HEVs as energy sources. There are vehicle-to-electricity grid technologies that could harness the energy stored in the HEV while not in operation and then feed it into the electrical grid. For HEVs the amount of electricity delivered by the battery equals the amount of electricity needed to recharge it and as a result, using HEVs as energy sources is feasible.

3.4 Socio-Economical Issues

In addition to improvements in air quality and fuel consumption there are social benefits associated with HEVs. For example they are seen as the "first realistic technological option for private transport that does not rely exclusively on gasoline" (Calef and Goble 2007). Other researchers discuss the changes in driving habits and attitudes that the use of HEVs and electric vehicles could bring. Brown (2001) reports that consumers driving hybrid electrical vehicles were more likely to plan their trips carefully reduce driving time and in general become more aware of the social implications of the transportation system. HEVs do not provide the functionality that would enable or permit inefficient or undesirable driving habits, such as "jump starts". Given that such behaviors are not desirable for any vehicle type, purchasing an HEV could be an opportunity to change driving habits toward safer and more responsible patterns and behaviors (Row 2009).

There are additional issues to consider in relation to HEVs. For example, the initial purchase cost of an HEV is higher than conventional vehicles, but the operating cost is much lower due to fuel consumption reduction. Salmasi (2007) also acknowledges that HEVs are the most economically viable solution, but also argues that savings of 70 % per gallon can be easily undermined by the amount spent on changing batteries, ultra capacitors, and so on, and as a result there is need to design a system that considers different energy portfolios. In addition, special consideration should be given to the design and development of drivetrain, control structures (which can be complex in HEVs), and vibration control in the vehicle (Salamasi 2007).

Maggeto and Mierlo (2001) state that the usage of the vehicle should be considered when "planning a suitable hybrid vehicle". The automotive purchase price is usually the baseline criteria for selecting a vehicle, but the reduced fuel consumption, emissions and changes in energy price should be taken into consideration: the "sticker price" therefore does not necessarily represent the real cost (Maggetto and Mierlo 2001).

It is also critical to consider economic issues associated with alternative transportation technologies. Granovskii et al. (2006) considered vehicle price, fuel cost and driving range as key economical variables in comparing different vehicle/fuel technologies. The vehicle price also included the additional cost for changing the batteries in hybrid and electrical vehicles. In terms of environmental factors, GHGs and air pollutants (APs) were included in the equation. Based on the analysis of the four vehicles compared (conventional gasoline ICE, HEV, EV and hydrogen FCV), the electricity generation scenario for fuel production impacts significantly the outcomes. If 50 % of the energy used to produce electricity comes from nuclear and renewable resources, then hybrid and electric cars become competitive. If however, fossil fuels account for more than 50 % of the energy sources, than hybrid cars are more advantageous than their electric, conventional and fuel cell counterparts (Granovskii et al. 2006).

The societal lifecycle cost (LCC) of several fuel/engine combinations was also investigated by Ogden et al. (2004). The authors consider the cost of vehicle and fuel, the cost for oil supply security and environmental costs due to GHG and air pollutant emissions. The vehicle/fuel options considered were compared against an advanced gasoline ICE vehicle that met Tier II air pollution standards. The options included:

- ICE vehicles fueled with hydrogen;
- ICE hybrid electric with gasoline, compressed natural gas, diesel, Fischer–Tropsch liquids, or hydrogen; and
- Fuel cell vehicles fueled with gasoline, methanol or hydrogen.

When costs for oil supply insecurity and environment (GHG and AP) are included in the overall lifecycle cost, all advanced options with the exception of FCV are less expensive than current gasoline SI ICE vehicle that is common today. The authors further report a damage cost from GHG emissions as $14–$510 per tonne of carbon as CO_2-equivalent. However, when the externality costs are not included, today's car has the lowest price. This reinforces the fact that environmental and fuel dependency factors when included would yield a reduced lifecycle vehicle cost. However, the problem is the value that society puts on such externalities.

3.5 Barriers to Hybrid Technology Adoption

Maclean and Lave (2003) used LCA to assess various vehicle and fuel options. They discovered that consumer acceptance can be a barrier to developing more environmentally friendly vehicles. For example, large vehicles have been supported by a large number of consumers and have slowed down the process of developing "greener" vehicles. Another issue is the contradictory nature of regulatory and societal goals. A smaller vehicle would satisfy sustainability principles developed by Anastas and Zimmerman (2003), however it could compromise safety and other regulations. Maclean and Lave (2003) suggest that the vehicle design and development stage is the most important one in creating sustainable vehicles. The vehicle operation (usage) stage contributes the most to GHGs in terms of CO_2 equivalents. Light duty vehicles that use diesel as fuel have an efficiency of 24 % compared to the gasoline ICE vehicles that have an efficiency of 20 %. As a result, diesel engine vehicles have the potential of higher fuel economy. Diesel has high carbon content, but because its production and vehicle efficiency are higher than for gasoline, it can reduce GHG emissions. The problem is that these vehicles have high NOx and PM emissions and are not highly sought by North American customers. HEVs achieve higher fuel economy and lower emissions, but on the other hand are more expensive and complicated in design. HEVs could become competitive with the conventional ICE through implementation of technological advancements that could increase fuel economy, lower emissions and vehicle initial price as well as increased social values assigned to GHGs or APs. Assigning a dollar value to environmental and social aspects is not always the preferred choice, however it does allow for comparisons between vehicle and fuel technologies. If all societal, economic, environmental factors as well as regulatory and customer goals are considered there is not one vehicle/fuel technology that is superior in all aspects (Maclean and Lave 2003).

Another issue to consider is the need for new infrastructure to deliver energy to vehicles using alternative powertrains. The authors report that if externality costs are valued low, then the advanced technologies cannot compare with the reference car unless the drivetrain does not cost more than the reference. In contrast, if the externalities are valued high then the advanced options would be competitive even if the cost of drivetrain is more than the reference car. In general, fuel-efficient liquid hydrocarbon fueled ICEVs and ICE-HEVs can achieve significant reduction in environmental and fuel uncertainty costs and also require minimum infrastructure changes. Even though the hydrogen fueled car yields the lowest cost when the externalities are valued high, the cost of implementing the infrastructure to deliver the hydrogen to consumers does not make it an attractive option in the near future (Odgen 2004).

4 Idle Reduction or Assistive Technologies for Fleet Vehicles

The availability of inexpensive petroleum based fuel has led the internal combustion engine (ICE) to dominate as the main propulsion system in motor vehicles. The convenience of petroleum fuel comes with a penalty principally in the form of tail pipe emissions. The operation of motor vehicles has become essential for transporting people and goods, but the drivetrain also powers onboard electronics and maintains the vehicle occupants comfortable. While engineers and designers optimize the ICE's efficiency to reduce fuel consumption and emission during operation, there is a trade-off in the ICE's idling efficiency. The ICE generates excessive power for idle conditions, wasting fuel and creating unnecessary emissions.

In 2007, the U.S. Department of Commerce's Vehicle and Inventory Survey estimated there are more than 400,000 commercial/transport trucks in service in the U.S. and each travels more than 500 miles a day (Lutsey et al. 2007). As illustrated in Table 2, the EPA estimates 960 million gallons of fuel are wasted per year from idling commercial trucks alone and the associated emissions include 180,000 tons of nitrogen oxide (NO_x), 5,000 tons of particular matter (PM), and 11 million tons of CO_2 (Frey and Kuo 2009). In response, a number of technologies for idling reduction (IR) have been pursued over the last two decades. Currently, the commercially available technologies are mainly for heavy transport truck. Passenger vehicles, however, have started to incorporate similar IR technologies. More generically, we can refer to these technologies as *assistive technologies* as they assist in the functioning of the vehicle services, and may in the future provide more functionality than simply reducing the impacts from idling.

IR technologies can be divided into two categories: onboard and wayside. Onboard technologies are installed on the vehicle itself to operate during idling. Wayside technologies are external infrastructure built to provide the necessary

Table 2 Long haul truck idling facts (U.S.) (adapted from Frey and Kuo 2009)

Vehicle and fuel statistics	
Number of trucks on the road	>400,000
Amount of diesel fuel used	960,000,000 (gallons)
Associated emissions	
NOx	180,000 (tons)
PM	5,000 (tons)
CO_2	11,000,000 (tons)

idling needs of the vehicles. Trucks have both onboard and wayside IR technologies available to them. They can be retrofitted or OEM equipped (Gains and Levinson 2009). Some technologies are mechanical modifications that reduce the fuel consumption of the ICE, while others function only during idling (Lim 2002).

4.1 Onboard Technologies

There are several onboard technologies that are used for idling reduction purposes. The most common types are:

4.1.1 Engine Start-Stop Control

This technology operates when a vehicle slows down to a complete stop. The onboard electronic control unit (ECU) then turns off the engine to prevent idling. The engine is restarted as soon as the brake pedal is release (automatic transmission) or when first gear is selected (manual transmission). An integrated starter-generator (ISG) is used instead of the conventional starter and alternator to provide the necessary engine cranking power without draining the battery.

4.1.2 Cylinder Deactivation

Under ideal conditions, fuel injection is ceased temporarily in designated cylinders of the engine to eliminate combustion. The 4-stroke cycle continues without combustion. Some systems open the exhaust valve(s) during the compression stroke and the intake valve(s) during the power stroke to minimize pumping loss. Cylinder deactivation is typically used on large displacement engines with six or more cylinders to operate temporarily the engine with only four cylinders firing to reduce fuel consumption. Uneven wear between the cylinders can occur however, and ECU controls can alternate the deactivated cylinders to prevent this.

4.1.3 Auxiliary Power Unit

An Auxiliary Power Unit (APU) is an external diesel powered generator installed into trucks. During extended idling, the main ICE is turn off and the APU is used to power the heating ventilation and air condition (HVAC) systems and all other accessories. Frey and Kuo (2009) reported that APU can reduce the fuel consumption by 36–47 %. Similar reductions were reported for SO_2 as well. In terms of other air pollutants it was found that NO_x emissions were reduced by 80–90 % while PM, CO, and HC emission reductions ranged from 10 to 50 %. However, APUs add extra weight, and cannot be used in 'creeping' conditions such as when slowly queuing at a border crossing (Frey and Kuo 2009).

4.1.4 Cab and Block Heater

During extended idling, waste heat recirculation from a small diesel heater is used instead of the ICE to heat the cabin and to maintain engine fluid temperature preventing cold weather engine start difficulties. Because no cooling can be provided, this configuration only works in cold weather conditions.

4.1.5 Air Conditioner

A battery powered air conditioning (A/C) system can be installed for use during extended idling when main ICE is turned off to save fuel. A secondary battery pack is charged during vehicle operation. Evaporative cooling and thermal storage are also available, but these additional functions can make the cost prohibitive. Because no heating can be provided, this configuration only works in hot weather.

4.2 Wayside Technologies

Wayside technologies are less common when compared to the onboard technologies. The most common types are: (1) single system; (2) dual system/shore power; and (3) fluid circulation systems.

4.2.1 Single System

Electrified parking spaces are built at rest stops or designated areas to provide HVAC for trucks when an extended stay is required. The charges depend on the user. Rest stop owners that install such wayside equipment can earn revenues by charging electricity usage (Gaines and Levinson 2009). However, the long term viability is still to be determined as this system is fairly new.

4.2.2 Dual System/Shore Power (SP)

This technology is similar to single system, but the electrified HVAC system is installed on the truck, which incurs an initial capital investment on drivers. The parking spaces need to have an electrical outlet built in, which translates on an increased capital investment for facility owners. Idling is eliminated and electricity cost is lower compared to diesel fuel (Gaines and Levinson 2009).

4.2.3 Fluid Circulation System

This technology is mainly used on buses at certain parking areas such school yards. The vehicle's coolant system is connected to an externally heating system and re-circulates the heated fluid to warm the bus during extended stays. However, this option requires significant capital investment and does not provide cooling and therefore is only effective in cool weather.

4.3 Applications to Fleet and Passenger Vehicles

Incorporating idling reduction systems or emerging assistive technologies on passenger vehicles is a recent development, most likely because of the increasing pressures to achieve sustainable modes of transportation. There are onboard IR technologies for non-commercial passenger vehicles (Stodolsky et al. 2001). However, the individual mobility of passenger vehicles works against wayside systems. Onboard technologies consist mostly of engine start-stop controls and cylinder deactivation battery systems. Other options being explored include battery systems. These may in fact become necessary as more traditional mechanical automotive systems (e.g., throttle control, power steering) become electrified.

Extended fuel consumption and air emissions are attributed to the unique operations of fleet vehicles and in particular, during idling. While drivers of passenger vehicles may have the option of simply not idling, fleet operators—and in particular emergency vehicle operators—may need to keep the vehicle operating to provide power to operate critical onboard equipment (e.g., computers, life saving equipment). These demands may be exacerbated during temperature extremes. However, prolonged idling can impose significant environmental and economic burdens. Hybrid vehicles have yet to be utilized widely by fleets, but there are other approaches to reduce emissions, including idling reduction technologies to operate in-vehicle equipment and maintain fleet vehicle capabilities instead of idling.

Overall, there are few studies published in terms of environmental and socio-economical impacts associated with the use of idling reduction technologies. However, IR or assistive technologies also share some of the advantages and disadvantages—albeit on a reduced scale—associated with hybrid electric vehicles because a number of them use conventional or extended batteries to reduce idling.

4.4 Driver Behavior

One important aspect to consider for the success of IR technologies is the driver "behavior" toward such technologies and how the driver interacts with the vehicle during routine or specialized activities. For example, if an IR technology is installed in a police vehicle, it should provide the necessary power to maintain laptop connectivity, emergency lights, and so on. Another example is maintaining the cabin temperature. The IR technology can be pre-programmed to keep the engine off until the cabin temperature drops/raises to a certain temperature depending on season. However, not all fleet operators have the same sensitivity toward temperatures. As a result, small but important considerations may influence significantly the drivers' attitudes towards adopting IR technologies.

4.5 Battery Recycling

Many types of batteries are available for automotive use; however, the lead acid battery (LAB) is currently the industry standard for automotive starting, lighting and ignition (SLI) as well as idling reduction (IR) technologies. Even with their low specific energy, LABs can withstand the automotive charge/discharge cycle better than other batteries and provide a high surge current. Its ease of construction also makes it ideal for mass production at low cost. However, should IR technologies be increasingly applied to fleet operations, there is significantly greater volume of batteries and different batteries that must be installed and eventually handled at their end-of-life, particularly when fleet vehicles turnover en masse.

LAB end-of-life (EoL) strategies have long been established and consist primarily of disposal (landfill) or recycling. Both strategies have their associated advantages and disadvantages in terms of the environmental impacts. According to Genaidy et al. (2008), LABs account for 88 % of the lead consumed in the U.S. with a 2.25 % increase in consumption each year. The disposal and recycling practices are therefore crucial in ensuring a sustainable life cycle of LABs. In addition to being sustainable, EoL strategies for LABs need to be economically feasible as well. Fisher et al. (2006) divided the financial costs of LAB EoL strategies into the following categories:

- Collection—this includes both labor and transportation costs. Collection cost is inversely proportional to volume. As volume increases, collection cost decreases.
- Sorting—a labor intensive step at local/regional waste facilities to separate the various types of batteries.
- Operation—operating procedures depend on the individual strategy and jurisdiction.

Collection and sorting cost can be assumed to be approximately the same for both disposal and recycling strategies. However, the operational costs are very different. The advantages and disadvantages of each EoL strategy can be evaluated by considering their economic feasibility and environmental impacts.

Strategy 1: *Disposal*

State-of-the-art landfill processes and controls better contain both air and water lead emissions. Processes and controls such as barrier layers in sanitary landfills can minimize heavy metal emissions to the environment (Fisher et al. 2006). With the increasing public opposition towards landfills and the growing scarcity of land due to population growth, the cost of landfills has been increasing steadily making this strategy less appealing. As a result, incineration has been growing in popularity, especially in high population density areas such as the European Union (EU). Incineration has an added benefit of energy recovery (combustion to electricity) that can be transferred back into the grid. However, the air emissions generated from the incineration processes have a negative impact on the environment and to a certain degree offset the benefits of this strategy. Both landfill and incineration have lower financial costs and have simpler logistics as compared to recycling.

Strategy 2: *Recycling*

LABs recycling is inherently energy intensive. Nevertheless, recycling and recovery rates of LABs have been high. The useful life of LABs is 4 years and the weight of lead content in each LAB is about 11 kg (Genaidy et al. 2008). It has been estimated that in 2003 about 2.6 million metric tons of lead are in the batteries of vehicles on the road (Environmental Defense 2003). Considering these values and the many vehicles on the road today, it is extremely important to consider recycling strategies that ensure the extraction of lead from LABs. The current practice is to pyrometallurgically extract metallic lead in rotary kilns. The breakdown of LAB components have been reported by Fisher et al. (2006) and are summarized in Table 3.

In the recycling process of LABs, H_2SO_4 (sulfuric acid), plastics and other materials are removed by combustion. However, this creates SO_2 and SO_3 that also contribute to acid rain and global warming gases such as NO_x (Volpe et al. 2009). The advantage of LAB recycling is to create a closed loop in the production life cycle, which in turn reduces the need for virgin lead. More research is necessary to

Table 3 Summary of LAB components (adapted from Fisher et al. 2006)

Component	%
Lead	65
Other metals	4
Sulfuric acid	16
Plastics	10
Other materials	5

increase the efficiency of current processes and to develop innovative technologies. At present, the costs of liquid fuel and electricity along with equipment and labor costs render recycling more costly in comparison to landfill disposal.

4.5.1 Environmental Assessment of LABs

Daniel and Pappis (2008) defined three environmental impact categories in their life cycle assessment of LABs: (1) resource consumption; (2) ecological impact; and (3) working environment impact. Apart from raw material consumption (lead, other metals, water, H_2SO_4, and plastic polymers), fuel and electricity are also consumed in the production of LAB. These impacts can be offset by recycling, which reduces the need for virgin raw materials. The main environmental impacts from all stages of the LABs lifecycle are summarized below.

- ***Global Warming/Greenhouse Effect***
  Measured in global warming potentials (GWP). All carbon emissions are converted to CO_2 equivalent in a 100-year time frame. For example, methane gas (CH_4) has 25 times the GWP as CO_2. For LABs, the stages that contribute the most to GWP are the collection (material) and distribution processes.
- ***Photochemical Ozone Formation***
  Caused by the release of volatile organic compounds (VOC) in the troposphere from the LABs life cycle. Similar to global warming, the production of VOCs is mainly from the collection and distribution processes.
- ***Acidification***
  Created from SO_2 emissions due to lead processing in the rotary furnaces after material collection and pryometallurgical recycling.
- ***Eutrophication***
  Occurs when nitrogen enters water bodies from transporting raw materials and the recovery/recycling processes.
- ***Eco-toxicity***
  Occurs mainly from the release of heavy metals and other hazardous materials into the environment. If LABs are disposed through landfill, then lead and acid are released into the soil. Air emissions would come from the incineration process. If recycling strategy is considered then lead, acid, and formaldehyde are released during pyrometallurgical processes.

The impacts from the production and assembly processes can be offset if large scale and efficient collection and recycling can be implemented (Bossche et al. 2006). However, many researchers regard this offset as a displacement of impacts, not as a true reduction of potential impacts. Nevertheless, without recycling and recovery, LAB life cycle will never become closed loop and eventual depletion of resources will occur.

The conventional disposal (landfill) of LABs promotes an open loop life cycle. Even with the added cost, recycling (recovery) of lead is essential to create a closed loop life cycle that will be sustainable. Since the initial 2004 EU Directive on batteries and accumulators, the battery industry has established the *Green Lead Vision* to make the industry as a whole more sustainable. Technological advancements in LABs recycling will continue to lower cost and environmental impacts. The move to hydrometallurgical LAB recycling by using cementation (reduction reactions) or electro-hydrometallurgical processes can lead to recovery rates as high as 99.7 % (Volpe et al. 2009).

5 Conclusion and Recommendations

Over the last four decades, there have been rapid developments in the hybrid vehicle technology, and in particular gasoline-electric powertrains. In addition ultra-clean diesel fuels have gained increasing interest. HEVs and clean fuels are seen as alternatives to conventional vehicles to conserve the natural resources and also protect the environment. There are several benefits attributed to the use of HEVs/clean fuel systems and particularly urban travel scenarios:

- Greater fuel efficiency than conventional gasoline cars;
- Substantial emission reduction;
- Reduced operating cost due to lower fuel consumption;
- Potential for the vehicle-to-grid technology to harness stored energy;
- Potential to enable changes in driving habits and attitudes; and
- Reduced health costs due to the improvement of air quality.

While these benefits cannot be realized in every operating scenario, HEVs can provide tangible economic and pollution control benefits in specific scenarios and furthermore, can provide potential significant and broadly-based environmental, health, and socio-economic benefits.

Fleet operations can benefit from hybrid technologies as well, but also from idling reduction technologies and assistive technologies. IR technologies have tremendous potential to reduce the impacts from idling. The immediate end users that will directly benefit from the outcomes of implementation of such technologies are fleet operators. The potential applications can be used by fleet operators as well as designers, engineers, non-governmental organizations, policy makers, and regulators to produce and commercialize vehicle technologies that will reduce environmental impacts from fleet operations. In particular, they may aid in establishing the rationale for IR technology suppliers and vendors as they promote the business case for their technologies. Finally, it may permit policy and decision makers to make more accurate decisions about the operation of fleet vehicles to reduce their environmental impacts while enhancing socio-economic benefits.

References

ACA (2008) Composites make sense for hybrid vehicles. Reinforced plastic, the American composites alliance

Albert JI, Kahrimanovic E, Emadi A (2004) Diesel sport utility vehicles with hybrid electric drive trains. IEEE Trans Veh Technol 53(4):1247–1256

Anastas PT, Zimmerman JB (2003) Design through the twelve principles of green engineering. Environ Sci Technol 37(5):94A–101A

Brown BM (2001) The civic shaping of technology: California's electric vehicle program. Sci Technol Hum Values 26(1):56–81

Burke A (2007) Batteries and ultracapacitors for electric, hybrid and fuel cell vehicles. Proc IEEE 95(4):806–820

Calef D, Goble R (2007) The allure of technology: how France and California promoted electric and hybrid vehicles to reduce urban air pollution. Policy Sci 40:1–34

D'Agosto M, Ribeiro SK (2004) Performance evaluation of hybrid-drive buses and potential fuel savings in Brazilian urban transit. Transportation 31:479–496

Daniel SE, Pappis CP (2008) Application of LCIA and comparison of different EOL scenarios: the case of used lead-acid batteries. Resour Conserv Recycl 52(208):883–895

Delgado RCOB, Araujo AS, Fernandes VJ Jr (2007) Properties of Brazilian gasoline mixed with hydrated ethanol for flex-fuel technology. Fuel Process Technol 88:365–368

Demirbas A, Demirbas MF (2011) Importance of algae oil as a source of biodiesel. Energy Convers Manage 52:163–170

Duoba M, Lohse-Busch H, Bohn T (2005). Investigating vehicle fuel economy robustness of conventional and hybrid electric vehicles. Paper presented at the EVS-21, Monaco, on, 2–6 April 2005

Easterbrook G (2000) Hybrid vigor. Atlantic Mon 286(5):16–18

Environment Canada (2010) Regulating Canada's on-road GHG emissions. www.ec.gc.ca. Accessed on Sept 2011

Environmental Defense (2003) Impacts of and alternatives for automotive lead uses: getting the lead out, report by environmental defense and the ecology center of ann arbor. Ecology Center, Clean Car Campaign, Michigan

Environmental protection agency (2008) Fuel economy guide. Available at: http://www.fueleconomy.gov. Accessed 2 Feb 2010

Fazal MA, Haseeb ASMA, Masjuki HH (2011) Biodiesel feasibility study: an evaluation of material compatibility; performance; emission and engine durability. Renew Sustain Energy Rev 15:1314–1324

Fisher K, Wallen E, Paul P, Collins L, Collins M (2006) Battery waste management life cycle assessment. Final report for publication. Environmental Resources Management

Frey HC, Kuo P-Y (2009) Real-world energy use and emission rates of idling long-haul trucks and selected idle reduction technologies. J Air Waste Manag Assoc 59:857–864

Gaines L, Levinson T (2009). Heavy vehicle idling reduction. Argonne National Laboratory, Illinois

Genaidy AM, Sequeira R, Tolaymat T, Rinder M (2008) An exploratory study of lead recovery in lead-acid battery lifecycle in US market: an evidence-based approach. Sci Total Environ 407(2008):7–22

Granovskii M, Dincer I, Rosen AM (2006) Economic and environmental comparison of conventional hybrid, electric and hydrogen fuel cell vehicles. J Power Sources 159:1186–1193

Honda World Wide, Automobile (2011) All new 2012 honda civic makes world debut at New York international auto show. http://world.honda.com/news/2011/4110420New-York-International-Auto-Show/index.html?from=r. Accessed 20 April 2011

ISO 14040 (2006) Environmental management—life cycle assessment—principles and framework. International standard, 2nd edn. www.iso.org

Kamimura A, Sauer IL (2008) The effects of flex fuel vehicles in the Brazilian light road transportation. Energy Policy 36:1574–1576
Karman D (2006) Life-cycle analysis of GHG emissions for CNG and diesel buses in Beijing. IEEE EIC climate change conference, Carleton University, Ottawa, pp 1–6
Katrasnik T (2007) Hybridization of powertrain and downsizing of IC engine—a way to reduce fuel consumption and pollution emissions—part 1. Energy Convers Manage 48:1411–1423
Lapuerta M, Armas O, Rodriguez-Fernandez J (2008) Effect of biodiesel fuels on diesel engine emissions. Prog Energy Combust Sci 34:198–223
Liaw YB, Dubarry M (2007) From driving cycle analysis to understanding battery performance in real-life electric hybrid vehicle operation. J Power Sources 174:76–88
Lim H (2002) Study of exhaust emissions from idling heavy-duty diesel trucks and commercially available idle-reducing devices. EPA 420-R-02-025
Lutsey N, Brodrick C-J, Lipman T (2007) Analysis of potential fuel consumption and emissions reductions from fuel cell auxiliary power units (APUs) in long-haul trucks. Energy 32:2428–2438
Maclean LH, Lave BL (2003) Life cycle assessment of automobile/fuel options. Environ Sci Technol 37:5445–5452
Maggetto G, Mierlo VJ (2001) Electric vehicles, hybrid electric vehicles and fuel cell electric vehicles: state of the art and perspectives. Annales de Chimie Science des Materiaux 26(4):9–26
Mastumoto S (2005) Advancement of hybrid vehicle technology. ISBN: 90-75815-08-5, EPE 2005, Dresden
Mizsey P, Newson E (2001) Comparison of different vehicle power trains. J Power Sources 102:205–209
Momoh OD, Omoigui MO (2009) An overview of hybrid electric vehicle technology. IEEE Xplore 978-1-4244-2601-0, pp 1286–1292
O'Dell J (2000) Part of the thinking behind hybrids: people may actually buy them. Los Angeles times, Feb 2, G1
Ogden MJ, Williams HR, Larson DE (2004) Societal lifecycle costs of cars with alternative fuels/engines. Energy Policy 32:7–27
Pourkhesalian AM, Shamekhi AH, Salimi F (2010) Alternative fuel and gasoline in an SI engine: a comparative study of performance and emission characteristic. Fuel 89:1056–1106
Rahman MA (2008) Advances on IPM technology for hybrid cars and impact in developing countries. In: 5th international conference on electrical and computer engineering ICECE, pp 20–22
Rambaldi L, Bocci E, Orecchini F (2010) Preliminary experimental evaluation of a four wheel motors, batteries plus ultracapacitors and series hybrid powertrain. Appl Energy 88:442–448
Row J (2009) The pros and cons of hybrid vehicle. http://www.edubook.com/the-pros-and-cons-of-hybrid-cars/4156/. Accessed 10 Feb 2010
Salmasi FR (2007) Control strategies for hybrid electric vehicles: evolution, classification, comparison, and future trends. IEEE Trans Veh Technol 56(5):2393–2404
Schouten NJ, Salman MA, Kheir NA (2003) Energy management strategies for parallel hybrid vehicles using fuzzy logic. Control Eng Pract 11:171–177
Stanciulescu V, Fleming SJ (2006) Life cycle assessment of transportation fuels and GHGenius. In: IEEE EIC climate change conference, Ottawa, pp 1–11
Stodolsky F, Gaines L, Vyas A (2001) Technology options to reduce truck idling. Transportation Technology R&D Centre, University of Chicago/Department of Energy (DOE)
Stone R (1999) Introduction to internal combustion engines 3rd edn. Society of Automotive Engineers Inc, SAE, Michigan, USA
Taylor GWR (2003) Review of the incidence energy use and costs of passenger vehicle idling. prepared for office of energy efficiency natural resources Canada. GW Taylor Consulting, Woodlawn
Turton H, Moura F (2008) Vehicle-to-grid systems for sustainable development: an integrated energy analysis. Technol Forecast Soc Chang 75:1091–1108

Uhrig ER (2006) Greenhouse gas emissions from gasoline, hybrid-electric and hydrogen-fueled vehicles. In: IEEE EIC climate change conference, pp 1–6

Van den Bossche P, Vergels F, Van Mierlo J, Matheys J, Van Autenboer W (2006) SUBAT: an assessment of sustainable battery technology. J Power Sources 162:913–919

Volpe M, Oliveri D, Ferrara G, Salvaggio M, Piazza S, Italiano S, Sunseri C (2009) Metallic lead recovery from lead–acid battery paste by urea acetate dissolution and cementation on iron. Hydrometallurgy 96(2009):123–131

Wehrey MC (2004) What's new with hybrid electric vehicles: recent advancements in battery technology and an overview of vehicle design, concept and prototypes. IEEE Power Energy mag 2(6):34–39

Part II
Tools and Methods for Greener Decision Making

Diagnostics Systems as a Tool to Reduce and Monitor Gas Emissions from Combustion Engines

Arkadiusz Rychlik and Malgorzata Jasiulewicz-Kaczmarek

Abstract Rapid increase in the number of vehicles, depletion of natural resources, pollution of the environment and the greenhouse effect require reduction of fossil fuel consumption and substitution of the product with alternative fuel as well as cutting down the emissions of hazardous and toxic substances generated by motor vehicles.

Keywords Diagnostic systems · Gas emission · Combustion engines

1 Introductory Information

1.1 Fuels

The source of power in motor vehicles is an engine. This creates a serious problem of engine selection in the aspect of its performance including the fuel issue. It is estimated that our civilisation has consumed energy equivalent to 500 billion tons of theoretical standard fuel, of which 2/3 in recent hundred years. It is anticipated that in this century energy consumption will rise by 250–300 % (Gronowicz 2004; Merkisz and Pielecha 2006).

As it results from different research, coal resources will be depleted at approximately 2200, crude oil at around 2050, and gas—approximately 2060. Therefore, intense works are ongoing globally on economical use of classical fuels

A. Rychlik (✉)
The University of Warmia and Mazury, Oczapowskiego 11, 10-719 Olsztyn, Poland
e-mail: rychter@uwm.edu.pl

M. Jasiulewicz-Kaczmarek
Poznan University of Technology, Strzelecka 11, 60-965 Poznań, Poland
e-mail: malgorzata.jasiulewicz-kaczmarek@put.poznan.pl

P. Golinska (ed.), *Environmental Issues in Automotive Industry*,
EcoProduction, DOI: 10.1007/978-3-642-23837-6_5,

(conventional), i.e. petrol, diesel oil, alternative fuels and use of renewable (infinite) energy resources.

Conventional energy is an energy using non-renewable (finite) resources, such as: hard coal, brown coal, natural gas or crude oil. Unconventional energy is an energy using renewable (infinite) resources such as: wind, water, sun, biomass, biogas, etc.

Alternative fuels (Niziński i inni 2011) are all fuels that are not products of crude oil processing (petrol, diesel oil) and which additionally (Jastrzębska 2007; Kneba and Makowski 2004):

- occur in sufficiently big quantities;
- are useful for car engine fuelling;
- are cheap in terms of production and sale;
- constitute a lesser risk for natural environment than traditional fuels;
- may be produced from own fossil materials and agricultural products;
- generate low costs of vehicle operation.

Considering the above facts it is advisable to examine the issues of current and future energy sources in the aspect of use thereof in fuel engines and to assess them in terms of diagnostics and generation of statuses given the gas emissions.

A low number of combustion engines with spark ignition are fuelled with liquefied petroleum gas (LPG) (0.6 MPa at $t = 20$ °C). The fuel is mainly a mixture of two hydrocarbons: propane C_3H_8 (18–55 %) and butane C_4H_{10} (45–77 %).

The main properties of this fuel and engines are as follows (Jastrzębska 2007; Kneba and Makowski 2004; Luft 2003; Merkisz and Pielecha 2006):

- density 1.5–2 times greater than that of air;
- volume consumption of gas is greater by 20–25 % than in the case of petrol fuelling due to lower calorific value with reference to the volume unit;
- LPG engines are fuelled with a stoichiometric mixture ($\lambda = 1$); they are equipped with a catalytic reducer;
- a slower combustion process causes the necessity of increasing of the ignition advance angle;
- with the same composition of the fuel–air mixture engines fuelled with liquefied petroleum gas are characterised by lower unit emissions of carbon monoxide (CO) and hydrocarbons (C_nH_m);
- nitric oxides (NO_x) emission comparing to petrol has yet not been unequivocally determined;
- particularly low exhaust emission is obtained for mono-fuel engines (Euro4 standard satisfied);
- total emission from bi-fuel vehicle fuelling systems is similar to mono-fuel fuelling;
- engines constitute the predominant solution for gas fuelling in passenger cars and lorries because of the system's low cost, light weight and dimensions.

There are five generations of LPG engine fuelling systems. Currently, generations IV and V are commonly used.

Fourth generation fuelling system (conforming to Euro 3, Euro 4 standards) is characterised by electromagnetic ejectors, which ensures reflection of the gas engine performance compared to petrol engine. In fifth generation systems (conforming to Euro 5 and Euro 6 standards) liquefied gas is not evaporated but injected as to the liquid phase to the engine's suction manifold. This results in engine power increase. No considerable differences have been observed between gas and petrol fuelled engines.

Natural gas (NG, methane CH_4—85–98 %) is natural fuel occurring independently or accompanying crude oil. Two forms of natural gas are used in combustion engines, namely:

- compressed (16–25 MPa—CNG—Compressed Natural Gas);
- liquefied ($t = -161.6$ °C—LNG—Liquefied Natural Gas).

The density of compressed natural gas is approximately four times lower than petrol's. Therefore, there is a requirement of using a high volume cylinder, which, as a consequence, increases the vehicle's weight. Gas fuelled engines are characterised by emissions at the level of Euro 4 standard. The gas is non-toxic, around two times lighter than air and easily mixes with air. Four generations of natural gas injection systems may be distinguished (Merkisz and Pielecha 2006).

The volume of liquefied natural gas (methane CH_4 ca 75 %) is approximately 600 lower comparing to its gaseous state. Thus, this property offers savings on transport and storing thereof. Storing of liquefied natural gas in the vehicle is performed using cryogenic tanks with powder-vacuum insulation. Liquefied natural gas fuelled engines meet the limitations of No_x; however, not for solid particles emissions. Application of LNG causes increase in energy storage comparing to CNG of the same volume.

Biogas is a gas generated in the process of anaerobic fermentation of manure and waste (mainly of agricultural origin) with the participation of methane bacteria (CH_4—58–85 %, CO_2—14–18 %). The fuel has properties similar to natural gas.

Animal waste (animal fat) may constitute a source of fuel for compression-ignition engines. Tests of the properties of such fuels, exhaust emissions, and engine performance are pending (Urząd Patentowy 2008).

Comparing to petrol, alcohols offer a series of advantages, to name just a few: theoretically lower demand of air for combustion, higher resistance to predenotation, lower ignition energy, higher flame velocity. The drawbacks include in particular: lower calorific value, difficulties with fuel evaporation, non-dissolving in petrol or diesel oil (a dissolvent, stabiliser or emulsifier shall be used), dissolving of certain metals, and acceleration of corrosion.

Methanol (CH_3OH) is made of natural gas; it is biodegradable but is a toxic compound. Low pressure of vapours is a cause of difficult starting of spark-ignition engines. Application of methanol results in increased engine power at the cost of increased engine wear. Application of methanol in compression-ignition engines is possible. Nevertheless, a foreign source of ignition or high level of compression is required.

Ethanol (C_2H_5OH) is obtained from plants (e.g. beet, cane, cereals, potato). The properties of ethanol are similar to those of methanol, except for toxicity and energetic density—higher by around 35 %.

Vegetable oils (biodiesels) that serve fuelling of compression-ignition engines include: rape, sunflower, soya, and palm oils and their esters. Two methods of vegetable oils use may be distinguished:

- combustion of fuels without preparation, by relevant adjustment of engine structures;
- preparation of fuels in the process of extraction (vegetable oil esters) as the basic method.

Due to their ability of spontaneous ignition, vegetable oils are more suitable for compression-ignition engines. They emit less vegetable black and, because of application of the catalytic converter, the emissions of CO, HC, and NO_x are also lower comparing to diesel engines. The drawback of esters is their aggressiveness toward certain plastics.

However, the central benefit of the use of vegetable oils is the closed CO_2 cycle in the atmosphere and, as a consequence, reduced greenhouse effect.

Once all the conventional fuels are depleted, the fundamental fuel for combustion engines and one of the basic factors supplying fuel cells will be hydrogen (Gronowicz 2004, 2008; Jastrzębska 2007; Merkisz and Pielecha 2006; Niziński i inni 2011). Hydrogen generated in the process of electrolysis of sea and ocean water may be transported and stored in gaseous, liquefied or chemically bound form in any place of the world. It should be pointed that hydrogen's main advantage as fuel is that in the combustion process reaction, water and small quantity of nitric oxides are obtained.

Nevertheless, basic future use of hydrogen will be supplying of a fuel cell that generates direct current that, after transformation into alternating current, may be used to power feed electric engines propelling motor vehicles.

It shall be emphasised that as a result of reaction of hydrogen and carbon monoxide dimethyl ether is obtained with a high cetane number ($CN > 55$) and high oxygen content, which indicates that the fuel may be used to power feed combustion engines with compression-ignition (Gronowicz 2008).

Ammonia is also a synthetic fuel obtained on a mass scale from hydrogen and atmospheric nitrogen. In normal conditions it is a gas; however, under pressure of 0.06–0.6 MPa it undergoes condensation and may be stored similarly to liquefied propane-butane. Application of NH_3 to power feed spark-ignition engines requires changes in the structure of combustion chamber and ignition system due to too high ignition temperature of mixture and low combustion velocity. Direct application of ammonia to fuel compression-ignition engines is impossible because of very difficult fuel spontaneous ignition. Therefore, a charge of diesel oil is used for it self-ignition or acetylene is added to the air–fuel mixture in the amount of 15–20 %.

Knowing the properties of alternative fuels for combustion engine fuelling and of the renewable fuel such as hydrogen, the status of know-how referring to engine diagnostics may be assessed and the development of their diagnostics systems, particularly in the context of natural environment preservation, may be anticipated.

1.2 Hazardous and Toxic Substances

Hazardous and toxic substances may be divided into two groups (Berhart et al. 1969; Günther 2002; Kneba and Makowski 2004; Luft 2003; Merkisz and Mazurek 2002; Merkisz and Pielecha 2006; Rokosch 2007; Rychter and Teodorczyk 2006):

- relating to engine, in particular classical one;
- non-engine related, including fuel vapours in fuel tanks, fine-particle dust caused by the use of brake lining and clutch facing as well as metal parts' corrosion, lead, copper, antimony, particles of erosion and catalytic coating of catalytic converters (platinum, palladium, rhodium).

Considering the purpose of this chapter, the authors will focus mainly on the aspect of emissions of gaseous pollutants relating to combustion engine operation.

Exhaust gases of combustion engines include hazardous and toxic substances (Fig. 1).

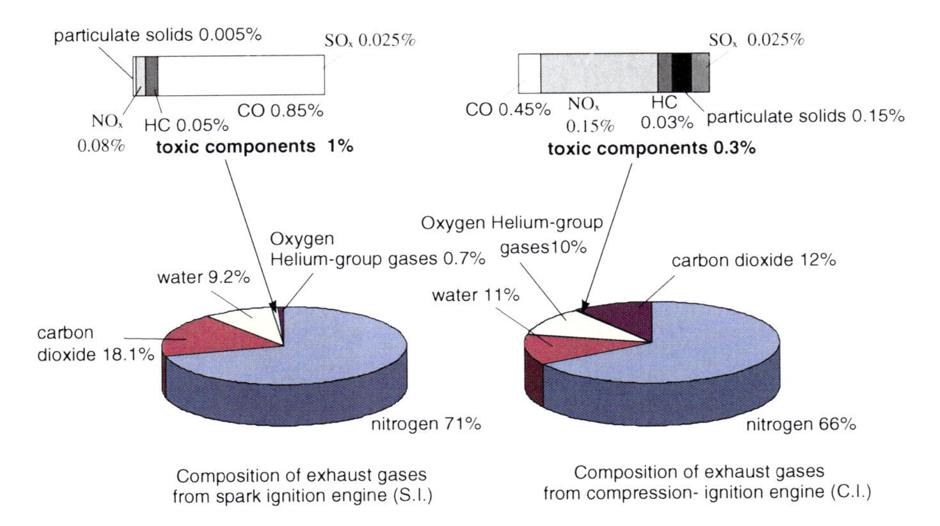

Fig. 1 Composition of S.I. and C.I. engine exhaust gases (Rokosch 2007)

Carbon monoxide CO is a colourless gas with no flavour, very hazardous for living organisms. Its concentration equal to 0.3 % causes death by suffocation. Incomplete combustion of fuel is the fundamental reason for carbon monoxide increase.

Hydrocarbons HC are, among others, the following chemical compounds: aliphatic hydrocarbons, alkanes i.e. chain (paraffin) hydrocarbons, cyclic aromatic hydrocarbons, aldehydes and multi-cycle hydrocarbons. They are cancirogenic. The causes of increased hydrocarbon emission are as follows:

- incomplete combustion of a rich mixture;
- local deficiency of oxygen in combustion chamber;
- combustion of a very weak mixture (valve burning, prolonged process of combustion);
- penetration of oil into the combustion chamber.

Nitric oxides NO_X are chemical combinations of N_2 nitrogen and O_2 oxygen, for instance: NO, NO_2, N_2O_5, formed in a temperature of over 1100 °C. Nitric oxides poisons human body, in particular nitrogen dioxide NO_2. Combination of nitric oxides and water H_2O form nitric acid HNO_3, nitrates HNO_2, that have a corrosive impact on engine elements.

High temperature of weak mixture combustion provides the most favourable conditions for nitric oxides generation.

Sulphur oxides SO_X are chemical combinations of O_2 oxygen and sulphur S_2 atoms included in fuel and water H_2O, formed in the process of fuel combustion. The results of these reactions are: SO_X, sulphurous acid H_2SO_3 and sulphuric acid H_2SO_4.

Sulphuric acid is the main reason of acid rains and, consequently, forest extinction. Sulphur dioxide SO_2 irritates the mucosa of eyes and nose, upper airway, and disturbs functioning of pulmonary alveoli.

Hydrogen sulphide H_2S may form in specific conditions in the catalytic converter. It causes poisoning showed in dyspnoea, loss of consciousness and breathing difficulties, irritation of the mucosa of eyes and airway.

Ammonia NH_3 may also form in specific conditions in the catalytic converter. It causes coughing, breathing difficulties, dyspnoea, skin redness and irritation or even loss of sight (direct contact).

Particulate matter PM (solids) is a product of incomplete combustion of hydrocarbons. These are clusters of carbon atoms—carbon black (particle nuclei—50 % made from: non-burnt fuel—6 %, engine lubricant—18 %, condensed water vapour, kinematic pair abrasive wear product, compounds of sulphur and ashes (26 %). They have different forms and sizes (0.05–15 μm), and merge in bigger clusters. They have a negative impact on upper airway and lungs.

Emission of particulate matter is approximately three times bigger in classical engines with compression-ignition comparing to spark-ignition engines. This does not apply to spark-ignition engines with direct fuel injection where the quantity is much bigger.

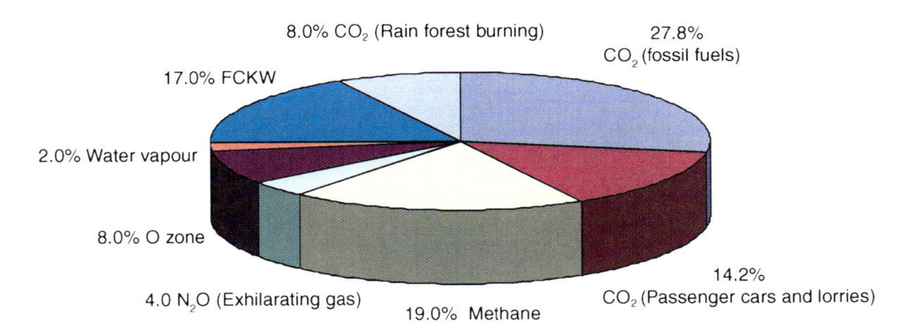

Fig. 2 Contribution of different gases to the Earth's greenhouse effect formation (Rokosch 2007)

Carbon dioxide CO_2 is a non-flammable gas, colourless, strongly toxic, naturally occurring and regulating life processes of plants and animals. Carbon dioxide is a product of complete combustion of fuel in the engine and a product of oxidation of toxic compounds in catalytic converter. Next to such gases as: methane CH_4, exhilarating gas N_2O, and halogens FCKW responsible for the greenhouse effect (Fig. 2).

1.3 Pollution limits

Since 1988 all newly registered vehicles in the USA have been equipped with an OBD (On-Board Diagnostic) as required by CARB (California Air Resources Board). In 1996, the enhanced version—OBD II—was introduced. The third version is OBD III which assumes abandoning of periodic inspections of vehicle elements and introduction of monitoring functions on current basis, i.e. informing superior units about failures and inadmissible exceeding of limit values for the exhaust emissions.

Since 2000 all spark-ignition engine vehicles newly registered in the European Union have been obligatory equipped with EOBD (European On-Board Diagnostic) and since 2004 the same requirement has been applied to compression-ignition engines (D-OBD—Diesel On-Board Diagnostic). Since 2006 also lorries have been covered by the obligation (Merkisz and Mazurek 2002; Rokosch 2007).

Introduction of an OBM (On-Board Measurement) for direct registration of toxic compounds is also planned. With OBD a crucial issue is to define the limits of pollution included in exhaust gases of combustion engines. The main institutions responsible for defining the limits are:

- in Europe—European Commission (EC);
- in the USA—Environmental Protection Agency (EPA); CARB (California Air Resources Board);
- in Japan—Ministry of Transport (MIT);
- in other countries—regulations included in European, American or Japanese standards.

Figures 3 and 4 illustrate changes of admissible exhaust emission levels for spark-ignition and compression-ignition engines for M class vehicles, determined in European standards.

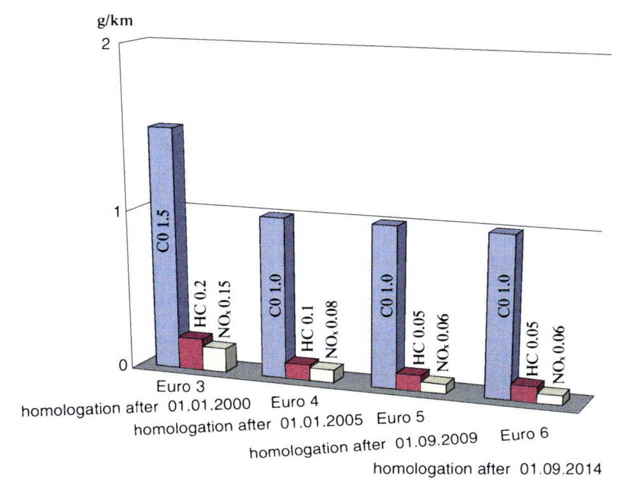

Fig. 3 European standards for the exhaust emissions for spark-ignition engine; M class vehicles; prepared based on (Rokosch 2007; Rozporządzenie (WE) NR 715/ 2007)

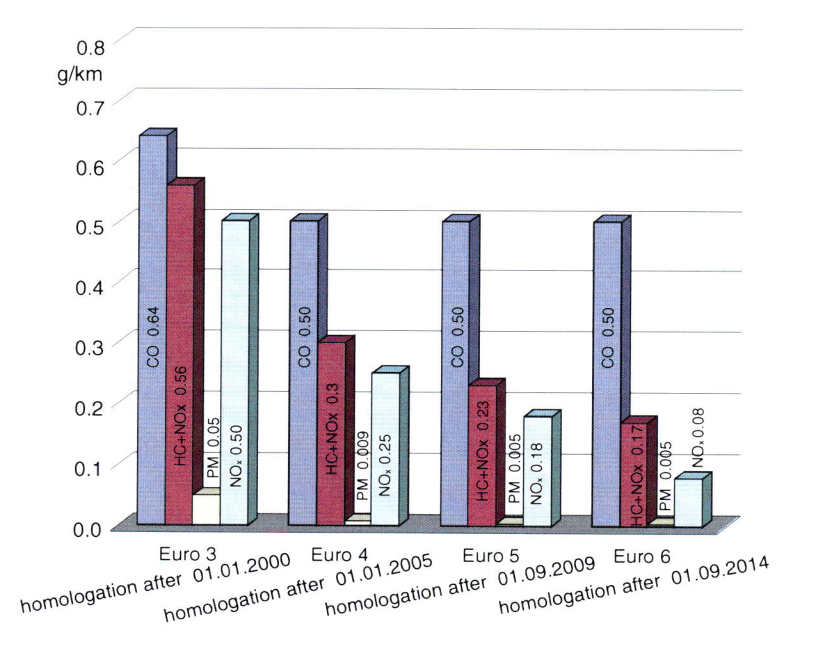

Fig. 4 European standards for the exhaust emissions for compression-ignition engine; M class vehicles; prepared based on(Rokosch 2007; Rozporządzenie (WE) NR 715/ 2007)

The European Automobile Manufacturers' Association (ACEA—French 'Association des Constructeurs Europeens d'Automobile') has assumed a voluntary obligation concerning reduction of fuel consumption in 2008 by approximately 25 % comparing to 1965. This means reduction of the average carbon dioxide CO_2 emission from 186 g/km to 140 g/km.

As a result of the global crisis the European Commission has slightly liberalized the dynamics of CO_2 emission reduction. According to the compromise that has been reached the average CO_2 emission in new vehicles manufactured between the years 2012–2015 shall drop to the level of 120 g/km. As target, by 2020 the average carbon dioxide emission is to reach 95 g/km (Fig. 5).

The emission of CO_2 is strictly connected with fuel consumption. Therefore, '115 g/km' virtually means the same as '4.4 dm^3/100 km'. Reducing of the carbon dioxide emission means nothing else than pressure to consume less fuel.

Analysis of the exhaust emissions of is carried out:

- during homologation-related tests;
- during periodic technical inspection of operated vehicles according to other measuring procedures, other than those mentioned above.

Homologation testing of new vehicles is carried out on an engine test bench in precisely determined engine loading conditions, which result from the assumed driving cycle, in specified sections.

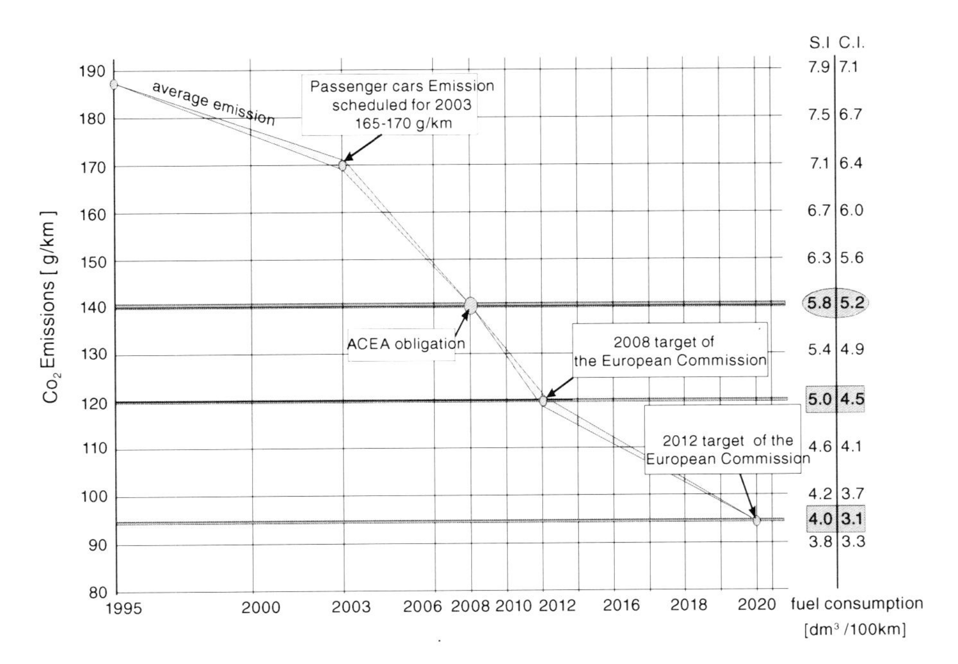

Fig. 5 Currently obtained and anticipated reduction of CO_2 emissions in Europe; prepared based on (Rokosch 2007; Rozporządzenie (WE) NR 715/ 2007)

In Europe it is NEDC that imitates typical traffic conditions (the cycle is composed of the urban segment (UDC) and extra urban segment (EUDC)). In other countries there are other driving cycles that apply (e.g. USA – FTP-75, Federal Test Procedure).

2 Properties of Motor Vehicle Diagnostic Systems

2.1 Motor Vehicle Diagnostic System

Due to the limited size of this chapter, only the on-board paradigms of motor vehicle diagnostic systems related to the monitoring of gaseous pollution emission will be discussed here. Readers interested in detailed issues shall refer to literature (Bocheński 2000; Bocheński and Janiszewski 1998; Günther 2002; Hebda et al. 1984; Herner and Riehl 2004; Janiszewski and Mavrantzas 2009; Merkisz and Mazurek 2002; Niziński 1999; Niziński and Michalkis 2002; Sitek 1999; Trzeciak 1998) where the methods and diagnostic devices used for motor vehicle diagnostics are described.

A structural approach to the **diagnostic system for a motor vehicle S_M** may be presented as follows (Fig. 6):

$$S_M = \langle S_p, S_z, R_{pz} \rangle \tag{1}$$

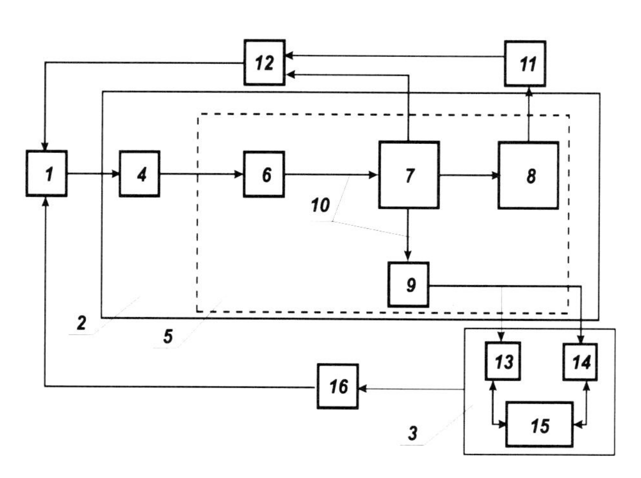

Fig. 6 A diagram of on-board, external diagnostic system for motor vehicles: *(1)* motor vehicle; (2) on-board diagnostic sub-system; *(3)* external diagnostic sub-system; *(4)* set of elements to be diagnosed; *(5)* communication sub-system; *(6)* set of sensors; *(7)* electronic control unit (ECU); *(8)* data display sub-system; *(9)* DLC; *(10)* data bus; *(11)* operator-driver; *(12)* actuator; *(13)* SAE handheld data reader; *(14)* generic data reader; *(15)* other diagnostic units; *(16)* analyst

where:
S_p on-board diagnostic sub-system;
S external diagnostic sub-system;
R_{pz} diagnostic relations

Principal functions of the on-board diagnostic systems include as follows:

- monitoring of the elements influencing the emission level of exhaust gases;
- recording of the information on malfunctions of the diagnosed elements;
- recording of the information on the operating conditions in which malfunctions occur;
- informing the operator about malfunctions that caused 50 % excess of the admissible emission of exhaust gases;
- forwarding the information to external diagnostic systems.

The on-board diagnostic system S_p is as follows:

$$S_p = \langle E, A, K, O, R_{EK} \rangle \tag{2}$$

where:
E a set of elements to be diagnosed;
A diagnostic algorithms;
K communication sub-system;
O software;
R_{EK} relations

The communication sub-system K may be described by:

$$K = \langle C, S, M, Z, Z_D, R_{CD} \rangle \tag{3}$$

where:
C a set of sensors;
S electronic control unit (ECU) – controller;
M data bus;
Z data display sub-system;
Z_D data link connector;
R_{CD} relations

External diagnostic sub-system $\mathbf{S_z}$ is described below:

$$S_z = \langle S_R, S_P, S_N, R_{RP} \rangle \tag{4}$$

where:
S_R generic (handheld) data reader;
S_p programmed (producer's) data reader;
S_N other diagnostic devices;
R_{RP} relations

As currently diagnostic systems (diagnostic monitors) do not form a centralised structure, sub-systems S_R and S_P form an onboard-external, i.e. mixed, diagnostic system of a motor vehicle. The fact is a result both of absence of a common standard for structure and communication of diagnostic sub-systems and different objectives assumed by the systems' manufacturers.

From the point of view of the effects of malfunction of diagnosed elements of motor vehicles we may assume the following classification (Rychter and Teodorczyk 2006:

(1) **emission-**related, causing:

- increased emission of toxic compounds (misfiring);
- increased noise (damage to the detonation combustion sensor);
- increased fuel consumption (damage to λ (lambda) probe).

(2) **driving safety-**related, having direct impact on vehicle's traffic safety. This applies mainly to systems such as: braking, steering, suspension, and lighting.
(3) **non-emission**-related, i.e. which does not increase exhaust emissions but deteriorate the vehicle's dynamics (damage to atmospheric pressure sensor).
(4) **driving comfort-**related, which deteriorate the driver's and passengers' comfort (damage to the brake pedal position sensor—cruise control activation impossible). These factors refer to the vehicle's body in particular.

The on-board sub-system of vehicle diagnostic systems (OBD) contains a definition of malfunction (Rychter and Teodorczyk 2006:

An unfit element (subassembly or function) shall be the one whose operation may cause considerable increase of exhaust gases emission or fuel consumption; where according to OBD II standard 'considerable' shall be defined as a growth by 50 % comparing to admissible value for this type of vehicle.

As it results from the malfunction definition the main task of OBD is current monitoring of the level of toxic compounds caused by engine failure or damage to other sub-assemblies. The Society of Automotive Engineers has determined emission standards included *inter alia* in the following norms:

- J1930—common terms and abbreviations to define elements critical in terms of emissions for all manufacturers selling their products in the USA,
- J1962—common data link connector (DLC) and its position in a vehicle,
- J1979—common reader for diagnostic data (SAE Scan Tool),
- J2190—diagnostic system functioning mode,
- J2012—common designation of unfitness (diagnostic trouble codes DTC),
- J1850—transmission protocol: computer—SAE Scan Tool.

In technical diagnostics, the basic terms are used, i.e. diagnostic algorithms which in OBD systems are called diagnostic tests and cover:

- tests of electrical fitness of measuring and actuating elements;

- passive tests of metrological fitness of actuating elements;
- functional tests of actuating elements;
- active tests of metrological fitness of measuring elements;
- emission tests.

Algorithms of electrical fitness of measuring and actuating elements refer to testing of: continuity of circuits, shorting of the reader's signal line or actuating feeding elements to the vehicle weight or feeding voltage.

Passive algorithm of fitness of measuring elements consists in testing of the correctness of sensor indication.

The algorithm of functionality of actuating elements consists in active testing of the elements' fitness using the known model signal.

The algorithm of active tests of metrological fitness of measuring elements is similar to the functional test. In this case, the measuring element is tested by forcing changes to the measured values by means of a controlling value that has an impact thereon.

The basis of OBD is **emission algorithm (test)** where the element fitness criterion has been defined as follows (Merkisz and Mazurek 2002)

An element or system is considered damaged if it alone (assuming that other elements or systems are fully fit) causes increased emission of HC, CO, NO_X above the limits admitted by standards. Both the emission values and the limits are expressed in g/km and are measured using the applicable test (in EU countries test I, in the USA FTP test). Otherwise, the element or system is considered fit.

According to OBD, the authors (Merkisz and Mazurek 2002) called the **emission tests (algorithms) monitors.**

Each monitor (algorithm, test) operates only one element or system influencing the emission. As per standards OBDII and EOBD, monitors are classified as follows:

- **continuous (unconditional)** that operate those elements which must be checked on current basis in any driving conditions and whose performance does not depend on other monitors' testing conditions;
- **non-continuous (conditional)** whose performance requires: completion of performance of other monitors and a longer observation time of parameter value change in determined driving conditions, i.e. a determined engine thermal and dynamic condition.

Forwarding of information in motor vehicles is made using **serial digital communication** enabling connection of elements and subassemblies in **functional nodes and integration of assemblies in one distributed control-and-measurement system.**

In OBD (1984, Fig. 7) the control-and-measurement system covered only the engine injection-ignition system. The controller of this system—ECM (Engine Control Module)—or ECM (Engine Management System)—controlled fuel

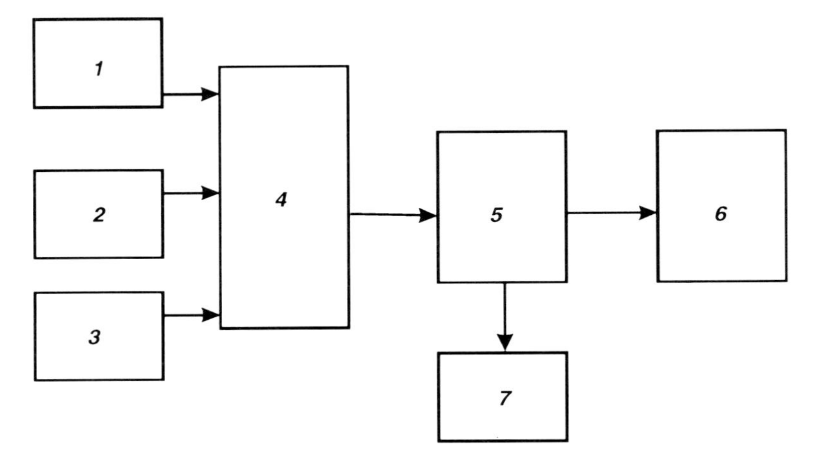

Fig. 7 Nature of the first generation on-board diagnostic system (OBDI). *(1)* engine injection system; *(2)* engine ignition system; *(3)* stabilisation system for idle run crankshaft rotational velocity; *(4)* sensor sub-system; *(5)* controller (ECM, EMS); *(6)* actuator; *(7)* diagnostic data reader

charge, ignition advance angle depending on loading, rotational velocity and engine temperature, all in real time.

Then, a digital connection was introduced to transmit data from ECM to external programming-and-diagnostic devices. Due to the development of vehicles, ECM was burdened with new tasks that resulted in its extension to PCM (Powertrain Control Module).

OBDII (1996) introduced decentralisation of the vehicle's control-and-measurement system by means of serial digital communication, connecting many independent elements and controllers operating different systems. Three types of digital communication were introduced (Merkisz and Pielecha 2006):

(1) **class A**—the purpose of this solution is to replace the classical electric beam with one transmission line between many devices forming the system's nodes. The A class communication is designed to transmit binary and continuous setpoints between the drivers and body equipment, in particular: switches, lights, air-conditioning, chassis, window lowering/lifting, etc. The maximum speed of data transmission is 10 kbit/s, and the average transmission time—100 ms;
(2) **class B**—serves to exchange digital data of the measured parameter values between controllers (nodes), eliminating the excessive number of results. The B class network must also perform the functions of the A class. The maximum speed of data transmission is 125 kbit/s, and the average transmission time—20 ms;

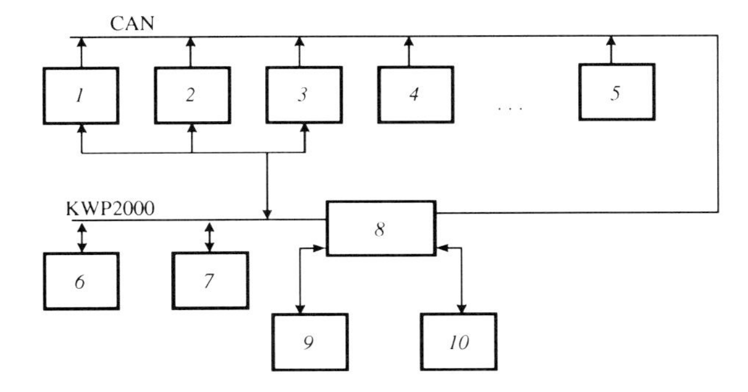

Fig. 8 Example of on-board network diagram (prepared based on (Merkisz and Mazurek 2002)). *(1)* engine controller (ECM); *(2)* transmission controller (TCM); *(3)* choke valve controller (ETC); *(4)* ABS controller; *(5)* other systems' controllers; *(6)*,*(7)* body controller; *(8)* data link connector (DLC); *(9)* SAE handheld generic reader; *(10)* programmed reader (producer)

(3) **class C**—enables transmission of the digital data of operating systems in real time, such as: ABS, ESP, gearbox. The C class network shall be compatible with A and B type networks. The maximum speed of data transmission is 1 Mbit/s, and the average transmission time—5 ms.

In the course of standardisation works (Ford, General Motors, SAE, Delphi, Bosch) several basic standards of digital communications were elaborated—B and C class, and particularly PWM, VPM, ISO, KWP2000 (Keyword Protocol—European B class system) as well as diagnostic communication standard with open architecture CAN (Control Area Network—BOSCH).

According to (Bocheński 2000 a typical structure of on-board network for B and C class vehicles will be CAN (Fig. 8).

Controllers from 1 to 5 require exchange of information in real time (C class). Therefore, they are connected via a data bus CAN (ISO 11898) which is also used to transmit data to the data connection link 8 and producer's reader 10. Controllers 1, 2, and 3 are responsible for the diagnostics of emission elements, thus, they were connected with a KWP (B class), which also serves data exchange between controller 6 and 7. This link is used to forward data to reader 9.

Connection of electronic systems of the vehicle with one or several data buses resulted in the following profits:

(1) reduced number of cables and, consequently:

- reduced weight of vehicle;
- increased reliability through reduced number of plugs and e.g. soldered connection;
- simplified run of cables, their installation and diagnostics;
- reduced cost of system production;
- reduction of problems related to electromagnetic compatibility.

(2) New possibilities of system connection via:

- better use of functioning possibilities;
- common strategy of regulation by different systems;
- multiple use of sensors thanks to which their number is lower and they are easier to control;
- flexible introduction of modernisation, possibly only new software without hardware changes (no new cables or controller adaptation);
- abandoning of small controllers.

(3) Improvement of diagnostic possibilities via:

- mutual controlling of systems;
- greater integration of system;
- detection of defects in the case of data transmission interruption.

(4) Relief for microchips, as no double transformation of signals in digital and the other way round occurs; data is exchanged digitally.

In CAN standard receiver addresses, also called identifiers, are transferred as the integral part of the transmission. Identifiers may be 11-bit or 29-bit. An identifier enables identification of the transmission contents (e.g. engine rotational velocity). The node (receiver) processes only the data whose identifier is included in the list of transmissions received by the node. All other data are ignored by the node. This method of addressing enables transmission of signals to many nodes, where the sensor sends its signals to the data bus directly or via the controller where they are distributed. In this way many variants of equipment may be performed easily, as further nodes may be connected to the already existing CAN system (open structure).

One of the CAN network's principles is that in the given moment many receivers may be active but only one transmitter. If CAN data bus is free each node may start data transmission. If many transmitters start data transmission simultaneously then the arbitration mechanism activates. The first transmitted data will be those of the highest priority, with no time and bit loss. Transmitters of data of lower priority send them automatically to receivers repeating the trials as long as the data bus is free again.

In OBDII information on defect that has occurred is displayed to the driver using MIL (Malfunction Indicator Light).

Generally, display of diagnostic information registered by the diagnostic system may be performed using: monitor, printer, digital or analogue indicators.

Vehicles are equipped with standard 16-plug DLC (Data Link Connector). This connector is placed in the cab on driver's side or in passenger seat area with access and identification possibility for service employees.

In standards SAE J1978 and ISO 150314 functional and electrical properties of a device to display and reading of information as well as interference on element status for OBDII have been determined. The standards include requirements concerning all solutions and manual tools, computer programmes and stationary diagnostic devices.

The device has been called **OBDII Generic Scan—GST Tool**. The basic GST functions include:

- automatic identification of serial transmission;
- display of diagnostic tests results;
- receipt and presentation of emission trouble codes;
- display of current values of parameters influencing exhaust gases emissions;
- deletion of error codes;
- other.
- The following OBDII scan tools are available:
- Handheld—generic;
- programmed (producer's).

A handheld reader (scan tool) (Fig. 9) is held in hands which facilitates analyst's work. There are two versions of handheld scan tools: basic and expanded.

In basic version, according to SAE, the following functions are possible:

- automatic detection of transmission protocol type;
- MIL status readout;
- diagnostic monitor readout;
- error code readout;
- readout of frozen frame;
- readout of parameter values;
- deletion (cleaning) of information.

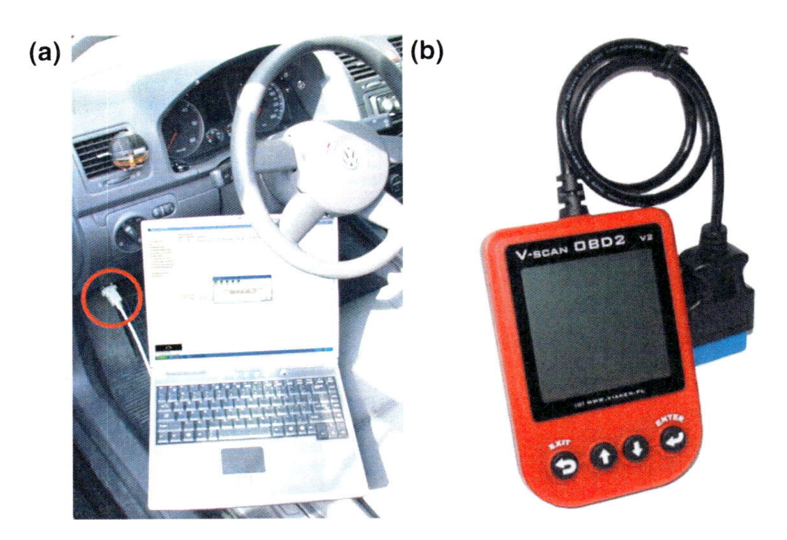

Fig. 9 General view of OBD data readers; (**a**) programmed scan tool; (**b**) handheld scan tool reader

Expanded versions of the handheld scan tool may include additional different information depending on the producer.

Programmed scan tool is a portable or stationary IBM/PC computer that enables the analysis of data on the status of elements obtained from OBDII or EOBD. Comparing to handheld scan tools this device has expanded functions of display of the results of filing, report printing, etc.

It shall be pointed that the external diagnostic system of a motor vehicle (expression 1, 4, Fig. 6) covers other diagnostic devices such as: exhaust gases analysers, engine braking, rolling devices, etc.

3 Diagnostics of Spark-Ignition Engines

3.1 Unconditional—Continuous Algorithm of the Combustion Process Diagnosis—Misfiring

The following is checked in spark-ignition engines:

- ignition system—using the engine's spot crankshaft rotational velocity or spot torque analysis;
- catalytic converter, oxygen sensors before and behind the converter, their heating system;
- systems: exhaust gases recirculation, fuel vapour evacuation, make-up air supply;
- elements of fuel and air feeding system, i.e. fuel pressure regulator, injectors, fuel tank and inlet system leakproofness;
- elements of adjustment the mixture composition in the function of rotational velocity and engine loading;
- air compressor and flowmeter;
- thermostat and timing gear system;
- electronic pedal of accelerator, brake and clutch;
- CAN data transmission network;
- other system, i.e. electric system circuits, sensors and actuating devices.

Depending on the exhaust emissions, spark-ignition engine elements may be divided into three groups:

(1) **Group A**—elements of the highest emission risk. A group includes elements and systems whose damaging or wear causes considerable increase of emissions. Vehicles operated with damaged elements of group A become significant emitters of toxic compounds. Group A covers the following elements: combustion process (detection of misfiring), catalytic converter, oxygen sensor, EVAP.

(2) **Group B**—elements of average emission risk. These include mainly special systems introduced to limit toxic compounds emissions: EGA exhaust gases recirculation system, make-up (secondary) air system, mixture ingredient control system, PCV (Positive Crankcase Ventilation).
(3) **Group C**—elements of potential emission risk. It covers system elements defined in American standards (none of which is covered by group A and B) which: directly or indirectly transmit the signal to central controller, or receive commands from the module, and damaging of which in normal operation conditions may cause increased emission, or are used in diagnostic procedures of other elements.

Algorithms of diagnostics of spark-ignition engine elements may be classified as:

(1) **unconditional**—which means that they may be performed in any driving conditions, combustion process diagnosis algorithm—misfiring;
(2) **conditional**, i.e. performance possible after obtaining positive results of other diagnostic tests.

Further in this chapter, the nature of several diagnostic algorithms will be discussed.

The following methods of identification of spark-ignition engine misfiring may be classified (Merkisz and Mazurek 2002; Rokosch 2007):

(1) analysis of spot value of the crankshaft angular velocity (Fig. 10);
(2) analysis of spot value of the exhaust gases pressure;
(3) measurement and analysis of the ionisation signal in combustion chamber;
(4) measurement and analysis of the engine's torque;
(5) analysis of optical signals recorded in the combustion chamber.

The analysis of the engine's spot crankshaft rotational velocity is commonly used at present (Fig. 10).

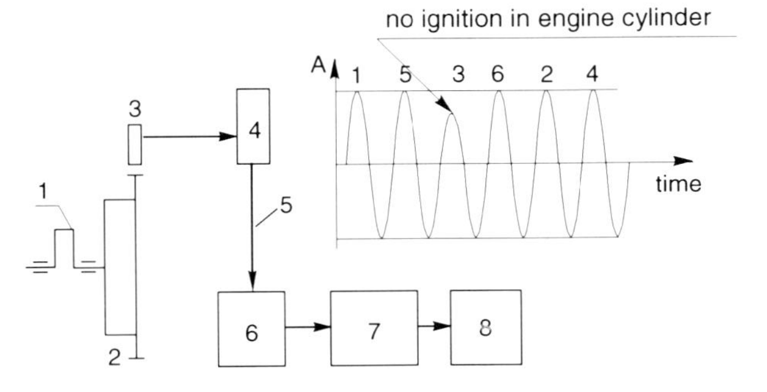

Fig. 10 Graphical presentation of a S.I. engine spot crankshaft rotational speed (Niziński i inni 2011). *(1)* crankshaft; *(2)* toothed ring; *(3)* inductive sensor; *(4)* signal matching system; *(5)* analogue signal; *(6)* A/C converter; *(7)* controller; *(8)* programme

3.2 Conditional—Non-continuous Algorithm of Catalytic Converter and λ Probe

Catalytic converter (catalyst) changes toxic compounds in exhaust gases into non-toxic. The following methods of catalytic exhaust gases treatment are known (Fig. 11) (Merkisz and Mazurek 2002; Rokosch 2007; Sterowanie silników o zapłonie iskrowym. Układy Motronic. Podzespoły. Informator techniczny BOSCH 2002; Sterowanie silników o zapłonie iskrowym. Zasada działania. Podzespoły. Informator techniczny BOSCH 2002):

(1) single **oxidising catalyst** (Fig. 11a);
(2) double **oxidising-reducing catalyst**—with make-up (secondary) air supply (Fig. 11b);
(3) single **oxidising catalyst with exhaust gases recirculation system** (Fig. 11c);
(4) single **multifunction catalyst** (Fig. 11d).

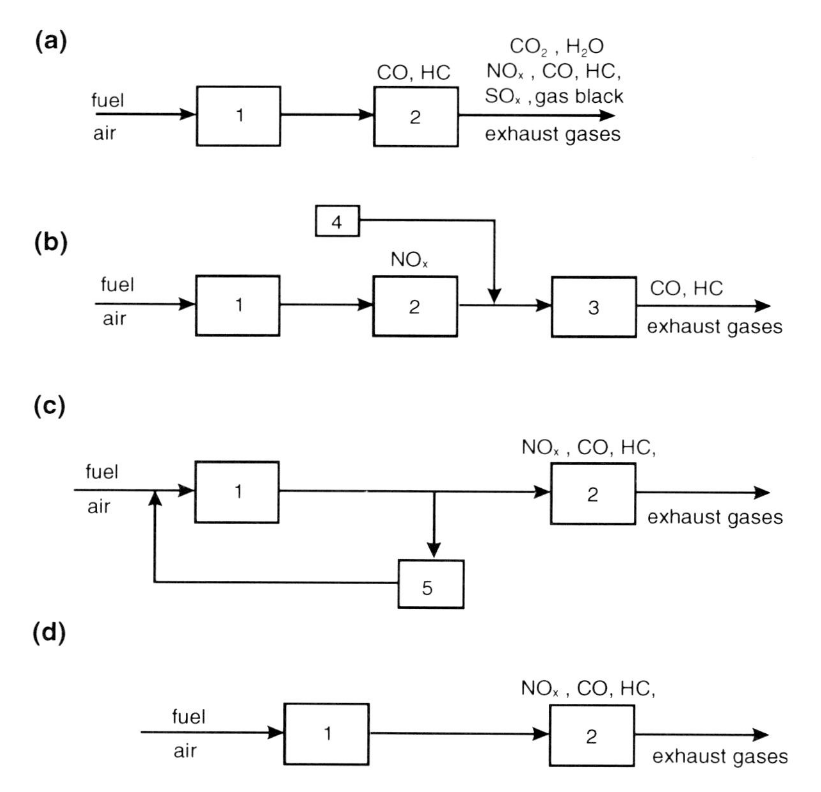

Fig. 11 A diagram of basic catalytic converters (described in text), (prepared based on (Merkisz and Mazurek 2002; Sterowanie silników o zapłonie iskrowym. Zasada działania. Podzespoły. Informator techniczny BOSCH 2002)); *(1)* engine; *(2)*, *(3)* catalytic converter; *(4)* make-up air valve; *(5)* exhaust gases recirculation valve

The task of oxidising catalyst (pre-converter, starting) is to combust carbon monoxide CO and hydrocarbons HC. The double oxidising-reducing catalyst reduces nitric oxides NO_x and oxidises CO and HC, while the single oxidising catalyst with exhaust gases recirculation valve enables reduction of NO_x, CO and HC.

Currently, catalytic converters of triple action are used where the following basic reactions take place:

$$2CO + O_2 \rightarrow O_2 \rightarrow 2CO_2 \tag{5}$$

$$2C_2H_6 + 7O_2 \rightarrow 4CO_2 + 6H_2O \tag{6}$$

$$2NO + 2CO \rightarrow N_2 + 2CO_2 \tag{7}$$

Euro 6 emission standard for M class vehicles propelled by spark-ignition engines includes the following limit values: CO—1.0 g/km; HC—0.10 g/km; NO_x—0.06 g/km; PM—0.005 g/km

The basic element of the converter is granulate carrier 1, most frequently ceramic granulate. Monolith, honeycomb structure is composed of three layers: catalytic, intermediate and ceramic carrier.

The catalytic layer enabling the reaction is a compound of platinum (base) and rhodium in 5:1 proportion. The intermediate layer contains caesium oxides as activators (promotors). Its task is to collect oxygen O_2 for $\lambda > 1$ and giving it for $\lambda < 1$ and increase calendar stability. The carrier's material is ceramics (cordierite) in the form of magnesium oxide MgO, aluminium oxide Al_2O_3 and silicone dioxide SiO_2. The number of canals per 1 cal^2 (cpsi) is 400–500.

The quality of catalyst performance may be described by the conversion coefficient **"k"**. It defines per cent conversion of individual toxic compounds in exhaust gases into non-toxic ones (Fig. 12).

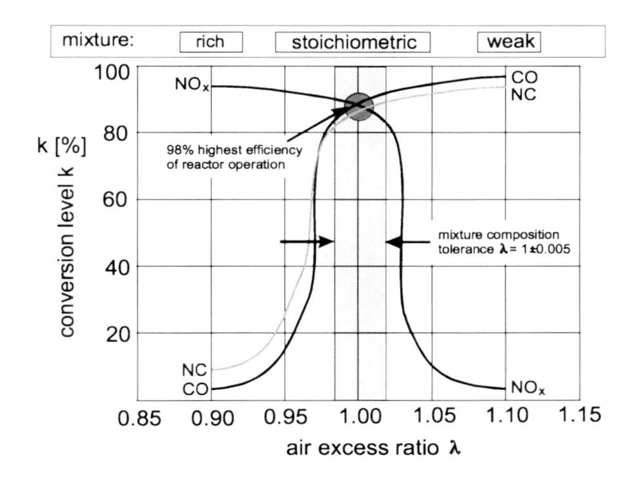

Fig. 12 Graphic presentation of the relation of conversion coefficient of a three-function catalytic converter of function λ, prepared based on (Rokosch 2007)

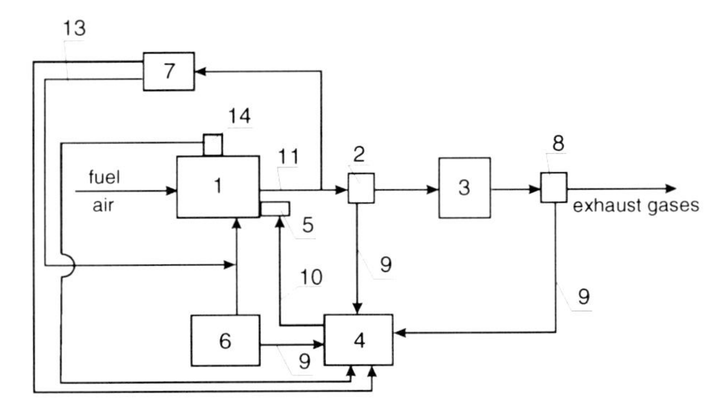

Fig. 13 A diagram of the system of adjustment of S.I. engine mixture composition (prepared based on (Sterowanie silników o zapłonie iskrowym. Układy Motronic. Podzespoły. Informator techniczny BOSCH 2002; Sterowanie silników o zapłonie iskrowym. Zasada działania. Podzespoły. Informator techniczny BOSCH 2002)): *(1)* S.I. engine; *(2)*, *(8)* oxygen sensor (probe λ); *(3)* catalyst; *(4)* controller; *(5)* injectors; *(6)* air filter (flowmeter); *(7)* exhaust gases recirculation valve; *(9)* measurement signal; *(10)* injector opening time control signal (fuel charge); *(11)*, *(12)*, *(13)* exhaust gases; *(14)* other sensors

$$k = \frac{i - j}{i} = 1 - \frac{j}{i} \times 100\,\% \tag{8}$$

where:

i concentration of exhaust gases before the catalyst;

j concentration of exhaust gases after the catalyst

It should be pointed that the catalyst is only one of the elements of the mixture adjustment system whose diagram is presented in Fig. 13.

Generally, the system functions as follows. Combustion engine 1 is fed with fuel and air that burns. Exhaust gases contain toxic components: CO, HC, NO_x.

The fundamental condition that must be absolutely satisfied to obtain major change in toxic exhaust gases into non-toxic is to supply combustion mixture of $\lambda = 1 \pm 0.005$.

The air excess ratio value λ is given by probes 2 and 8 which send a signal to controller 4. The controller also collects necessary signals from air flowmeter 6 and other sensors, such as: rotational velocity of engine crankshaft, temperature of cooling liquid, air temperature, loading and other. Based on the collected data the controller sends a controlling signal to rejectors 5, in order to determine their opening time, i.e. define the fuel charge $D = f(t_w)$, and, consequently, to determine λ, as per the engine operating conditions.

As it has been previously mentioned, adjustment of the mixture composition in spark-ignition engine is possible because of signals obtained from oxygen sensors (probe λ).

Air sensor is composed of a ceramic insert made of zirconium dioxide, stabilised with yttrium oxide, and Mg–Si oxide-coated porous platinum electrodes.

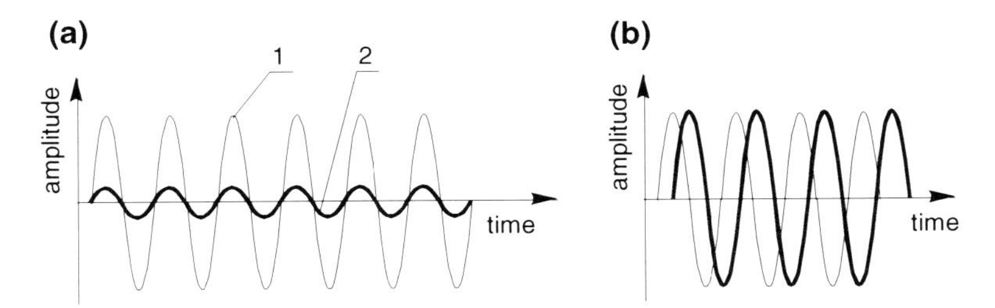

Fig. 14 Graphic presentation of the condition of catalytic converter (prepared based on (Rokosch 2007)) (**a**) fit catalyst; (**b**) unfit catalyst, *(1)* sensor signal after catalyst; *(2)* sensor signal after catalyst

The flow of exhaust gases in the exhaust pipe flushes probe λ, the inside of which is connected with atmosphere 8 (ca. 21 % O_2). Engine exhaust gases contain a certain number of oxygen particles O_2. Therefore, a potential difference occurs on probe electrodes, which causes U_s voltage current flow in the voltage indicator circuit.

The value of signal transmitted by probe λ depends on the type of fuel–air mixture supplied to the cylinder (rich, weak).

Diagnosing of spark-ignition engine catalytic converter is performed based on the data contained in the signals received from probe λ (Fig. 13). If the catalyst is fit there are considerable differences between the signals (Fig. 14a) and if the catalyst is damaged the signal from sensor located after the catalyst is no longer significantly weaker and is similar to the signal from the sensor located before the catalyst (Fig. 14b).

The air sensor enables precise adjustment of the mixture composition critically influencing the value of exhaust gases emission. Its diagnostics consists in monitoring of the voltage of measuring element and measurement of voltage and current in the sensor heating circuit. Figure 15a illustrates the course of a fit sensor signal and a damaged sensor signal, while Fig. 15b shows a signal from a new and a worn sensor. Figure 15c and d presents deceleration of the sensor's reaction to modification of the mixture composition resulting from its aging.

3.3 Algorithm of Diagnostics of Fuel Vapours Evaporation System

The purpose of the evaporative emission control system (Evaporation Prevention—EVAP) (Fig. 16) is to reduce the emission of hydrocarbons resulting from evaporation and escape to atmosphere of the light fractions of fuel in the tank.

Due to high temperature, vapours of fuel in tank 3 flow through the pre-valve 5 to the filter via absorber 6. The absorber's insert shall be activated carbon with diversified surface structure (200–1500 m^2/g), having perfect absorption features.

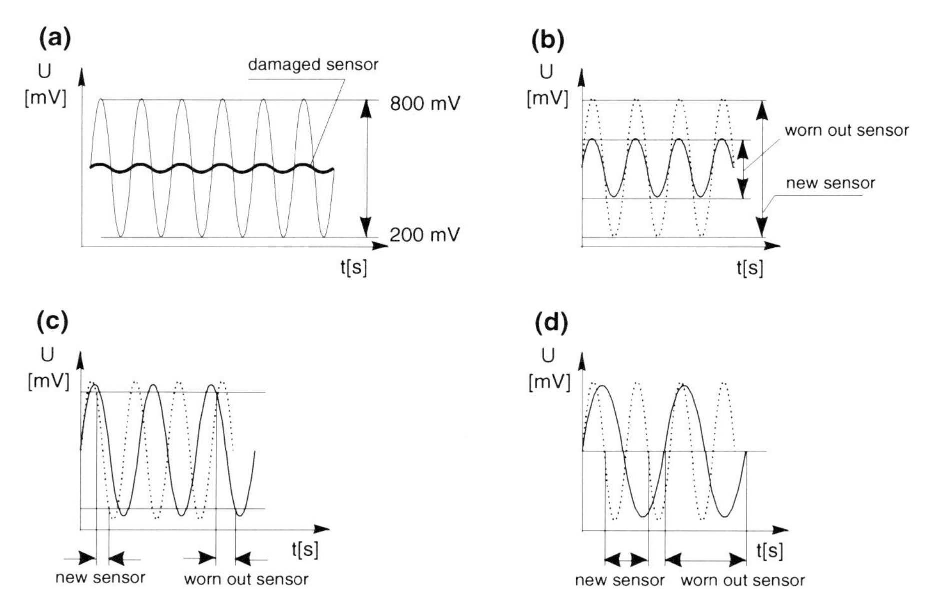

Fig. 15 Changes of oxygen sensor signal (probe λ) (Janiszewski and Mavrantzas (2009: (**a**) fit and damaged sensor; (**b**) new and worn-out sensor; (**c**) verification of sensor reaction; (**d**) prolongation of sensor reaction time

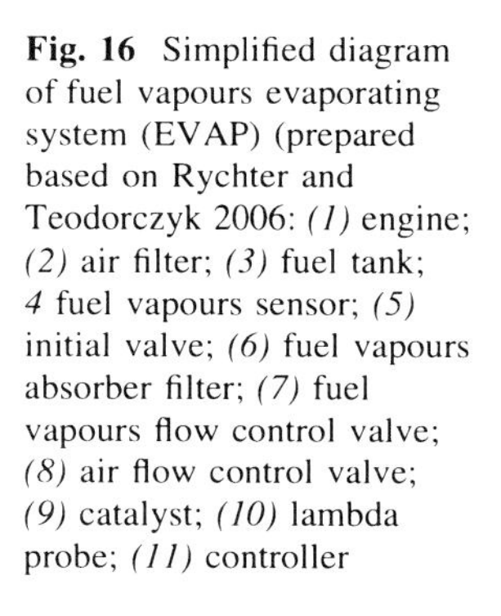
Fig. 16 Simplified diagram of fuel vapours evaporating system (EVAP) (prepared based on Rychter and Teodorczyk 2006: *(1)* engine; *(2)* air filter; *(3)* fuel tank; *4* fuel vapours sensor; *(5)* initial valve; *(6)* fuel vapours absorber filter; *(7)* fuel vapours flow control valve; *(8)* air flow control valve; *(9)* catalyst; *(10)* lambda probe; *(11)* controller

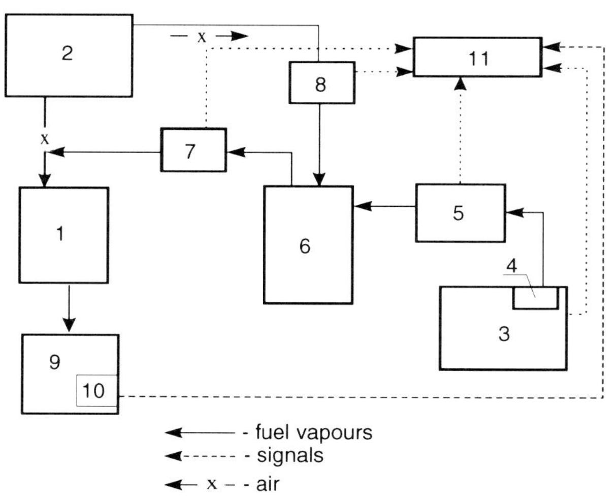

Opening of control 7 causes that a part of air drawn in by the engine flows first by absorber 6 catching the fuel vapour and carrying it off to the engine's suction manifold where they are burnt. The statuses of the system's operation are controlled by controller 11 with the use of the signal from probe λ 10.

Depending on the level of absorber filling, three operational statuses of EVAP system may be distinguished:

(1) empty absorber—the system works, the engine feeding mixture is weaker;
(2) overfilled absorber—the system works, the mixture in enriched;
(3) absorber filling conforms to mixture $\lambda = 1$.

The criterion that determines the system's unfitness is leakage. In OBDII, EVAP is monitored by the system of leakage detection VLDS II (Second Generation Vacuum Leak Detection System). The diagnostic signal is decay of negative pressure in the system compensated with the opening diagram: $\emptyset = 1$; 0.5; 0.25 mm.

4 Diagnostics of a Compression-Ignition Engine

4.1 Diagnostics of Exhaust Gases Recirculation System

Only some of the tests justified in the case of spark-ignition engines may be, after relevant modification, applied to compression-ignition engines. Below presented are the tests applied exclusively to compression-ignition engines concerning:

- injection advance angle;
- fuel injection initiation;
- fuel charge values in the rotational velocity and engine loading function;
- misfiring;
- status of: injectors, exhaust gases recirculation system, glow plugs, fuel pressure accumulator, particle matter filter, catalytic converter and compressor;
- status of controller and CAN data bus;
- sensors of, for instance: engine rotational velocity, make-up air temperature, injector needle travel, exhaust gases temperature, etc.

Considering the high number of elements to be diagnosed, in general, only some of the diagnostic procedures will be discussed here. Namely, those concerning exhaust gases recirculation, particle matter filter, and catalytic converters.

In the case of compression-ignition engines, there are two methods to reduce toxic exhaust gases emissions, i.e. by (Merkisz and Mazurek 2002; Rokosch 2007; Sterowanie silników o zapłonie samoczynnym Podzespoły. Informator techniczny BOSCH 2000):

- optimisation of the fuel combustion process;
- use of particle matter filter and catalytic converters.

As it has been described above, favourable conditions for engine operation that contribute to generation of high concentrations of nitric oxides NO_x occur with

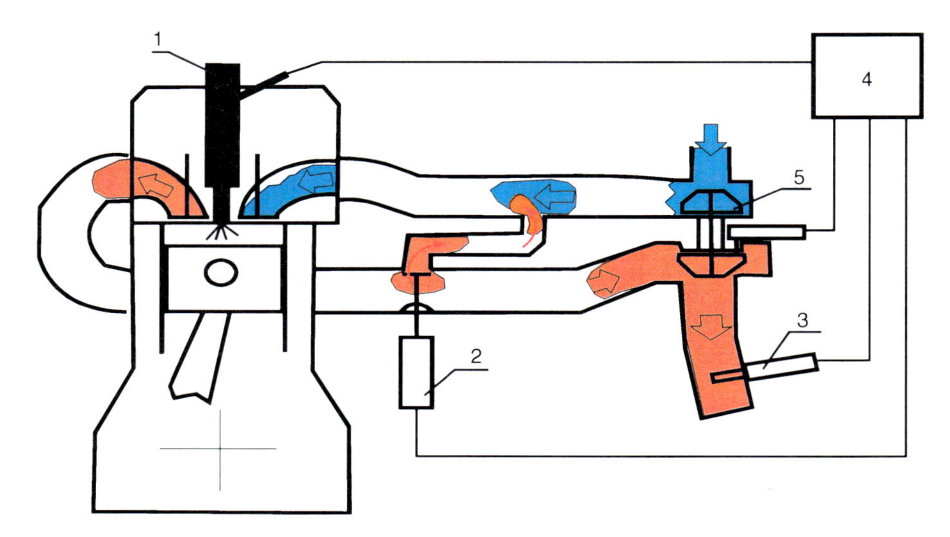

Fig. 17 Nature of exhaust gases recirculation system operation; *(1)* fuel injector, *(2)* exhaust gases recirculation system valve EGR, *(3)* λ probe, *(4)* EDC controller, *(5)* turbo compressor

high combustion temperatures of weak mixtures ($\lambda < 1$). An important system fostering the reduction of nitric oxides is the **exhaust gases recirculation system**.

The system directs a part of the exhaust gases (0–30 %) from the exhaust manifold to the suction manifold via an EGR valve (Exhaust Gas Recirculation) (Fig. 17).

Supplying the exhaust gases to the combustion chamber results in the reduction of spot values, local combustion temperatures of oxides, thus reducing the intensity of binding between nitrogen N_2 and oxygen O_2, i.e. reducing the quantity of nitric oxides (NO, NO_2, N_2O, N_2O_5).

Application of exhaust gases recirculation at the value of 0–15 % does not cause any growth of the CO emission. While with a higher level of exhaust gases recirculation the combustion time extends, making the emissions of CO, HC, and PM go up.

The recirculation of exhaust gases proves to be a good solution both for S.I. and C.I. engines. It allows reducing the emissions of NO_x even up to 50 %.

Direct measurement of CO, HC, NO_x, PM emissions is possible only with a very complex measuring system. Therefore, OBD system diagnoses emissions indirectly by identifying the statuses and parameters of:

- closing and opening of the EGR valve;
- exhaust gases recirculation valve;
- atmospheric air pressure sensor;
- mass air flow sensor.

It shall be here pointed that a reduction of toxicity of compression-ignition engine exhaust gases may be obtained in the following ways (Rokosch 2007):

- high-pressure fuel injection;
- more intense air whirling in the combustion chamber;
- shortening of injection time;
- injection of the guiding charge;
- four-valve technology (four valves per cylinder);
- Common-Rail system and injection unit.

4.2 Diagnostics of Particle Matter Filter

Particle matter filter is an effective tool used to limit the emission of solid particles (PM—see item 1.1) outside the C.I. engine. The filtering insert is made of cordierite, titanium (Ti) and aluminium (Al) compounds, metals with porous foamy structure or sintered metal oxides. A part of the canals through which the exhaust gases flow ($\varnothing = 0.0001$ mm) are 230–300 cpsi (cell per square inch). Filter fitness $\eta \approx 99$ %.

During engine operation the filter is polluted with solid particles. Therefore, its diagnostics and regeneration becomes a crucial issue.

The filter of solid particles may be regenerated by burning the carbon black and hydrocarbons with one of the following methods (Rokosch 2007):

(1) **without fuel additives;**
(2) **with fuel additives.**

The first method covers two solutions:

(a) artificial increasing of exhaust gases temperature by adequate improvement of the fuel injection process, using fuel late-injection (Fig. 18a) at the exhaust stroke. The result of late-injection is that fuel does not combust in the combustion chamber but only evaporates. Fuel combustion takes place only in the particle matter filter, which reduces their value even by 40 %. It must be emphasised that combustion of carbon and hydrocarbons included in the PM takes place at $t > 600$ °C, i.e. it is an active filter regeneration (Fig. 18b).
(b) burning of solid particles using a catalyst that lowers the minimum combustion temperature to 350–500 °C, at which the solid particles are burnt (Fig. 18b) (passive regeneration). The catalyst replaces additives supplied with fuel.

Application of fuel additives enables lowering of the minimum temperature necessary to regenerate the filter to 400 °C. The additives are caesium or ferrum oxides.

Diagnostics of a polluted solid particles filter consists in measurement of pressure value before and the filter, as such pressure is much higher. Additional performance of exhaust gases temperature measurement and using of the air

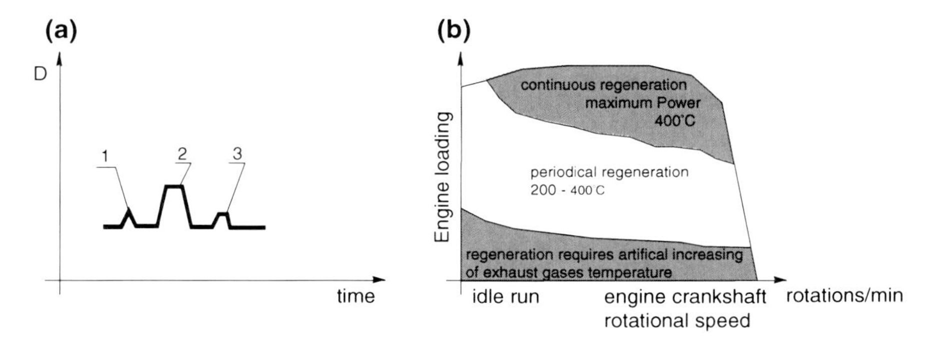

Fig. 18 Illustration of late injection (**a**) and conditions of engine operation where regeneration of the particular solids filter supported with catalyst is possible, *(1)* fuel charge; *(2)* supply charge (basic); *(3)* fuel late injection; prepared based on (Rokosch 2007)

flowmeter signal allows assessing the filter's pollution and initialising of the regeneration process.

4.3 Diagnostics of Catalytic Converter

Compression-ignition engines combust weak mixtures ($\lambda > 1$). Hence, the emissions of CO and C_mH_n are lower than in the case of spark-ignition engines. Nevertheless, the problem of lowering the emission of NO_x still remains. Exhaust gases recirculation is not sufficient for the purpose.

Two major methods of NO_x reduction is exhaust gases may be distinguished (Rokosch 2007; Sterowanie silników o zapłonie samoczynnym Podzespoły. Informator techniczny BOSCH 2000):

- **with accumulating catalyst of NO_x**
- **selective catalytic reduction—SCR.**

The nature of accumulating NO_x reduction is based on the following reactions (Fig. 19):

$$2NO + O_2 \rightarrow 2NO_2 \tag{9}$$

$$2BaO + 4NO_2 \rightarrow 2Ba(NO_3)_2 \tag{10}$$

$$Ba(NO_3)_2 + 3CO \rightarrow 3CO_2 + BaO + 2NO \tag{11}$$

$$2NO + 2CO \rightarrow N_2 + 3CO_2 \tag{12}$$

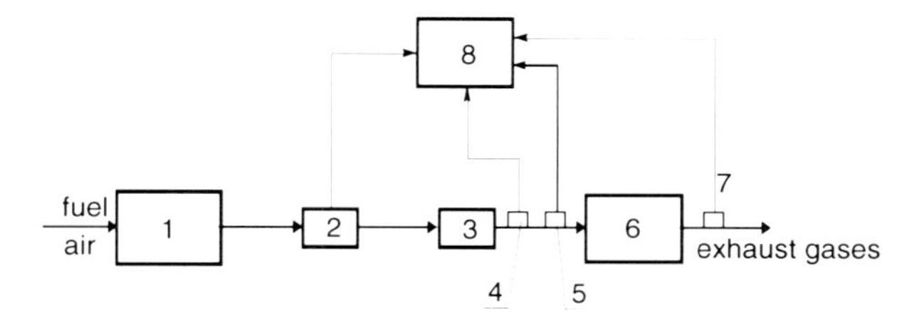

Fig. 19 A diagram of NO_x recirculation system accumulating method (prepared based on (Rokosch 2007; Sterowanie silników o zapłonie samoczynnym Podzespoły. Informator techniczny BOSCH 2000)), *(1)* engine; *(2)* exhaust gases heating (option); *(3)* oxidizing catalyst (option); *(4)* temperature sensor; *(5)* broadband lambda probe LSU; *(6)* accumulating catalyst NO_x; *(7)* NO_x sensor or lambda probe; *(8)* controller

The exhaust gases heater 2 is to reduce the emission of toxic compounds when starting the engine. A reaction (1.19) takes place in the catalyst. The process of NO_x reduction is made of phases: accumulation, discharge and conversion.

The accumulation phase (storing) (60 s of engine operation—$\lambda > 1$) nitric oxides are oxidised on the platinum layer to nitrogen dioxide NO_2. Then, NO_2 reacts with special oxides at the surface of the catalyst forming nitrates with oxygen. For instance, as a result of NO_2 reaction with barium oxide BaO, barium nitrate $(NO_3)_2$ is formed (12). The accumulating NO_x catalyst accumulates nitric oxides generated in the case of an engine operating in the excess of air.

The discharge phase covers regeneration of the catalyst, i.e. the accumulated nitric oxides must be processes (1.13). For that reason the engine is switched to operate on a rich mixture ($\lambda < 1$–2 s. of engine operation). Here, the reducing substances are H_2, CH, and CO.

In conversion phase, reaction (1.13) takes place where the rhodium coat finally reduces NO, in the presence of CO, to N_2 and CO_2.

It shall be pointed that the broadband λ probe (German 'Lambda—Sonde Universal') is able to measure the air excess ratio with broad limits ($\lambda = 0.7$–3.4); hence, it is used for petrol, gas, and diesel engines. The probe's typical property is its linearity.

The SCR method assumes dosing of 32.5 % diesel exhaust fluid $(NH_2)_2CO$), from which ammonia (NH_3) is obtained in catalyst 6 (Fig. 20).

The fluid is supplied using pump 12, from tank 11, to injector 4 located in the exhaust manifold of the engine.

The following reaction takes place in oxidizing catalyst 3:

$$2NO + O_2 \rightarrow NO_2 \tag{13}$$

while in catalyst 6 the reactions of:

(a) thermolysis:

$$(HN_2)_2CO \rightarrow NH_3 + HNCO \tag{14}$$

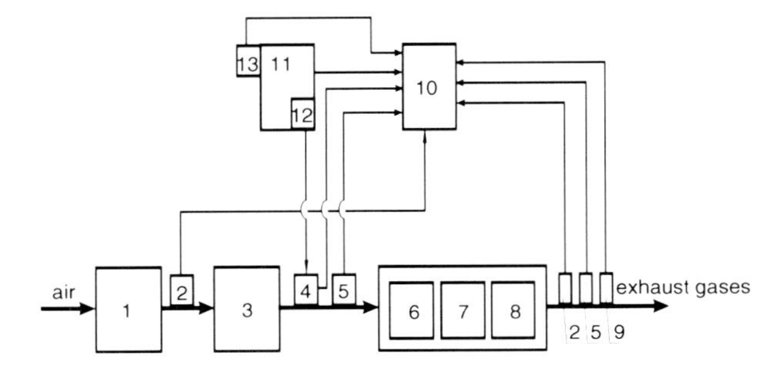

Fig. 20 A diagram of NO_x reduction system in C.I. engines using SCR (prepared based on (Sterowanie silników o zapłonie samoczynnym Podzespoły. Informator techniczny BOSCH 2000)): *(1)* engine; *(2)* temperature sensor: *(3)* oxidizing catalyst; *(4)* injectors; *(5)* NO_x sensor; *(6)* hydrolytic catalyst; *(7)* SCR catalyst; *(8)* NH_3 blocking catalyst; *(9)* NH_3 sensor; *(10)* controller; *(11)* reducing agent tank; *(12)* pump; *(13)* reducing agent level sensor

(b) hydrolysis:

$$HNCO + H_2O \rightarrow NH_3 + CO_2 \tag{15}$$

Reduction of nitric oxides to N_2 and H_2O takes place in catalyst 7 according to the following reaction:

$$NO + NO_2 + 2NH_3 \rightarrow 2H_2 + 3H_2O \tag{16}$$

Catalyst 8 prevents NH_3 discharge to the atmosphere. The conversion of nitric oxides approximates 90 %. Additionally, the system enables reducing of the emission of particle matter by circa 40 %.

SCR diagnostics is based on the measurement of:

- the level of reducing agent in the tank (by measuring of the solution's electric conduction).
- fluid temperature, when the agent freezes at −11 °C its volume grows by 10 %, while exceeding the temperature of 40 °C accelerates chemical degradation of the urea solution;
- electric conduction of urea allowing detection of fluid dissolving or application of incorrect fluid.

Data from sensors 2, 4, 5, 9, 11, 13 (Fig. 20) are supplied to the controller (diagnostic system) where they are processed to make the diagnosis on fitness or unfitness (control light on) of the system.

5 Trends in Motor Vehicle Diagnostic Systems Improvement

Currently applied OBD systems (Merkisz and Mazurek 2002; Rokosch 2007):

- are good tools for early detection of single malfunctions of elements relating to exhaust gases emission; they enable repairing of contemporary more and more complex engine control systems;
- remember the engine operating conditions, which enables malfunction location;
- offer a possibility to detect incorrect signals from sensors, damaged servos and other actuating segments, which enables the engine service-repair process;
- the OBD possibilities are limited as they do not make automatic identification of the causes of malfunction possible;
- do not always enable detection of complex consequential malfunction;
- do not give grounds to foresee the consequences of malfunction;
- do not offer direct monitoring of concentrations of toxic compounds of exhaust gases;

The trends to improve diagnostic systems for motor vehicle engines are determined by numerous factors, in particular:

(1) lowering of the admissible limits of toxic compound emission and inclusion or so far not limited exhaust gases in the toxicity assessment;
(2) influencing the toxic exhaust gases emission by:

- modification of engine structure that offers to most favourable combustion process, optimisation of the opening and valve travel phases as well as change of the compression level;
- application of new systems of charge combustion combining the properties of circulation of S.I. and C.I. engines, which eliminates the differences between the engines;
- more frequent use of supercharged S.I. engines;
- application of two smaller pressure charging units instead of one compressor;
- application of new catalytic converters, filters and other elements.

(3) considering a compromise between the ecological and economical requirements of the purposeless of designing sophisticated engine element solutions;

- changing the type of fuel, which generates structural modification of engines allowing for the following criteria: ecological, fuel consumption reduction, vehicle movement dynamics and driving comfort.

Given the analysed issues, the development of motor vehicle diagnostic systems will head the following directions:

- increased durability of the solutions applied, aiming at limitation of toxic exhaust gases emissions;

- development of new methods to monitor ecological properties of combustion engines;
- construction of new generation sensors for measurement of measured values (physical quantities);
- application of sensors to measure concentrations of toxic compounds in exhaust gases;
- direct measurement of concentrations of gaseous components of exhaust: CO, C_nH_m, NO_x and particle matter;
- monitoring of the process of combustion in engines and optimisation of fuel injection parameters using new diagnostic signals;
- replacing OBD with OBM (On-Board Measurement) system, enabling direct measurement of exhaust gases concentration values which will ensure more precise control over the emission properties of engines even if consequential malfunction occurs;
- extending the OBD obligation onto motorcycles;
- introducing exhaust gasses limits and OBD for engines of other machinery and stationary engines;
- development of a diagnostic method to assess the quality of fuel used;
- development of hierarchical diagnostic models and diagnostic algorithms for entire motor vehicles;
- development of on-board concepts and onboard-external diagnostic system concepts for entire motor vehicles;
- development of diagnostic methods and algorithms as well as diagnostic systems for engines fuelled with hydrogen;
- development of methods and algorithms to diagnose fuel cells;
- development of hierarchical diagnostic models and diagnostic algorithms as well as diagnostic systems for hybrid vehicles and electrically propelled vehicles.

Given the analysed issues, it has been found as follows:

- motor vehicle diagnostics faces serious challenges caused by introduction of new, alternative fuels, especially LPG, natural gas: CNG and LNG, vegetable oils, biogas, alcohols and hydrogen;
- because of determining the levels of toxic compounds in exhaust gases, currently used OBD meet the basic requirements of the monitoring of status and malfunction location in S.I. and C.I. engines;
- there is a trend to substitute OBD with a system of direct measurement of toxic levels (OBM) that will enable more precise definition of the physical properties;
- there is a need of extending the toxic vapours elimination systems onto all vehicles, including military motor vehicles;
- it is necessary to elaborate improved OBDs covering the entire motor vehicle;
- diagnostics of hydrogen fuelled engines, fuel cells, hybrid and electrical vehicles need to be introduced;
- systems must be provided to control safety of motor vehicle.

To summarise the analysed issues concerning formal aspects and assessment of civil diagnostic systems for motor vehicles the following may be stated:

- a motor vehicle diagnostic system consists of two sub-systems: on-board and external;
- with reference to consequences, malfunction of diagnosed elements may be classified as: emission-related, safety-related, non-emission related, driving comfort-related;
- to diagnose motor vehicle elements, relevant algorithms are used, including but not limited to: electric fitness of measuring and actuating elements, emission—continuous—unconditional and non-continuous—conditional;
- in contemporary motor vehicles data is transferred using a serial digital communication of classes A, B, C;
- a key element of OBD is the communication sub-system including: sensors, controller, data display panel, DCL and software;
- spark-ignition combustion engines are equipped with well-developed algorithms of diagnosing the elements influencing the level of exhaust gases toxicity, particularly: misfiring, catalytic converter, oxygen sensor (λ probe), exhaust vapours evaporation system, fuel charge defining.
- compression-ignition engines are also equipped high diagnostic quality solutions for such elements and systems as: exhaust gases recirculation system, particle matter filter, catalytic converter.

References

Berhart M, Dobrzyński S, Loth E (1969) Silniki samochodowe. WKŁ, Warszawa

Bocheński C (ed) (2000) Badania kontrolne samochodów. WKŁ, Warszawa

Bocheński C, Janiszewski T (1998) Diagnostyka silników wysokoprężnych. WKŁ, Warszawa

Gronowicz J (2004) Ochrona środowiska w transporcie lądowym. ITE, Radom – Poznań

Gronowicz J (2008) Niekonwencjonalne źródła energii. ITE, Radom–Poznań

Günther H (2002) Diagnozowanie silników wysokoprężnych. WKŁ, Warszawa

Hebda M, Niziński S, Pelc H (1984) Podstawy diagnostyki pojazdów mechanicznych. WKŁ, Warszawa

Herner A, Riehl HJ (2004) Elektrotechnika i elektronika w pojazdach samochodowych. WKŁ, Warszawa

Janiszewski T, Mavrantzas S (2009) Elektroniczne układu wtryskowe silników wysokoprężnych. WKE, Warszawa

Jastrzębska G (2007) Odnawialne źródła energii i pojazdy proekologiczne. WNT, Warszawa

Kneba Z, Makowski S (2004) Zasilanie i sterowanie silników. WKiŁ, Warszawa

Luft S (2003) Podstawy budowy silników. WKiŁ, Warszawa

Merkisz J, Mazurek S (2002) Pokładowe systemy diagnostyczne pojazdów samochodowych. WKŁ, Warszawa

Merkisz J, Pielecha I (2006) Alternatywne napędy pojazdów. Politechnika Poznańska, Poznań

Niziński S (ed) (1999) Diagnostyka samochodów osobowych i ciężarowych. Bellona, Warszawa

Niziński i inni (2011) Systemy diagnostyczne wojskowych pojazdów mechanicznych. ITE, Radom

Niziński S, Michalkis R (2002) Diagnostyka obiektów technicznych. ITE, Radom
Rokosch U (2007) Układy oczyszczania spalin i pokładowe systemy diagnostyczne samochodów OBD. WKŁ, Warszawa
Rozporządzenie (WE) NR 715/2007-W sprawie homologacji typu pojazdów silnikowych w odniesieniu do emisji zanieczyszczeń pochodzących z lekkich pojazdów pasażerskich i użytkowych (Euro 5 i Euro 6) oraz w sprawie dostępu do informacji dotyczących naprawy i utrzymania pojazdów
Rychter T, Teodorczyk A (2006) Teoria silników tłokowych. WKŁ, Warszawa
Sitek K (1999) Diagnostyka samochodowa. Wyd. Auto, Warszawa
Sterowanie silników o zapłonie iskrowym. Układy Motronic. Podzespoły. Informator techniczny BOSCH, WKŁ, Warszawa (2002)
Sterowanie silników o zapłonie iskrowym. Zasada działania. Podzespoły. Informator techniczny BOSCH, WKŁ, Warszawa (2002)
Sterowanie silników o zapłonie samoczynnym Podzespoły. Informator techniczny BOSCH, WKŁ, Warszawa (2000)
Trzeciak K (1998) Diagnostyka samochodów osobowych. WKŁ, Warszawa
Patent application No. P-386761: Sposób i instalacja do utylizacji odpadów w postaci tłuszczy zwierzęcych. Urząd Patentowy (2008)

Reachability of Multimodal Processes Cyclic Steady States Space

Grzegorz Bocewicz

Abstract Cyclic scheduling problems are usually observed in FMSs producing multi-type parts where the AGVS plays a role of a material handling system as well as in other various multimodal transportation systems where goods and/or passenger itinerary planning plays a pivotal role. Schedulability analysis of concurrently flowing cyclic processes (SCCP) executed in these kind of systems can be considered using a declarative modeling framework. Consequently, the considered SCCP scheduling problem can be seen as a constraint satisfaction one. Assumed representation provides a unified way for performance evaluation of local cyclic as well as supported by them multimodal processes. The main question regards of a control procedure (e.g. a set of dispatching rules) guaranteeing a SCCP cyclic behavior. In this context, the sufficient conditions guaranteeing both local and multimodal processes schedulability are discussed and some recursive approach to their designing is proposed.

Keywords Cyclic processes · Declarative modeling · Constraints programming · State space · Periodicity · Dispatching rules

1 Introduction

Operations in cyclic processes are executed along sequences that repeat an indefinite number of times. In everyday practice they arise in different application domains (such as manufacturing, time-sharing of processors in embedded systems, digital signal processing, and in compilers for scheduling loop operations for parallel or pipelined architectures) as well as service domains (covering such areas

G. Bocewicz (✉)
Department of Electronics and Computer Science, Koszalin University of Technology, Koszalin, Poland
e-mail: bocewicz@ie.tu.koszalin.pl

P. Golinska (ed.), *Environmental Issues in Automotive Industry*, EcoProduction, DOI: 10.1007/978-3-642-23837-6_6,

as workforce scheduling (e.g., shift scheduling, crew scheduling), timetabling (e.g., train timetabling, aircraft routing and scheduling), and reservations (e.g., reservations with or without slack, assigning classes to rooms) (Dang et al. 2011; Gaujal et al. 1995; Liebchen and Möhring 2002; Song and Lee 1998; Steger-Jensen et al. 2011; Trouillet et al. 2007; Wójcik 2007; Wang et al. 2007; Pinedo 2005). Such cyclic scheduling problems belong to decision problems, i.e. aimed at searching for answering whether a solution possessing the assumed features exists or not (Trouillet et al. 2007). Moreover because of their integer domains the problems considered belong to a class of Diophantine problems (Guy 1994; Smart Nigiel 1998); hence some classes of cyclic scheduling problems can be seen as non-decidable (undecidable) ones (Bocewicz et al. 2009b).

Therefore, taking into account non decidability of Diophantine problems one can easily realize that not all the behaviors (including cyclic ones, i.e. represented by cyclic schedules corresponding to cyclic steady states of the system) are reachable under constraints imposed by system's structure. The similar observation concerns the system's behavior that can be achieved in systems possessing specific structural constraints. That means, since system's constraints determine its behavior, hence both system structure configuration and desired cyclic schedule have to be considered simultaneously. So, cyclic scheduling problem solution requires that the system structure configuration must be determined for the purpose of processes scheduling, yet scheduling must be done to devise the system configuration.

In that context, our contribution provides discussion of some solvability issues concerning cyclic processes dispatching problems, especially the conditions guaranteeing solvability of the cyclic processes scheduling problem. Their examination may replace exhaustive and time consuming searching for solutions satisfying required system functioning.

Many models and methods have been proposed to solve the cyclic scheduling problem (Levner et al. 2010). Among them, the mathematical programming approach (usually IP and MIP (Dang et al. 2011; Von Kampmeyer 2006)), max-plus algebra (Polak et al. 2004), constraint logic programming (Bocewicz et al. 2009a, 2011a, b; Bocewicz and Banaszak 2013; Wójcik 2007) evolutionary algorithms (Cai and Li 2000) and Petri nets (Song and Lee 1998) frameworks belong to the most frequently used. Majority of them are oriented at finding of a minimal cycle or maximal throughput while assuming deadlock-free processes flow. The approaches trying to estimate the cycle time from cyclic processes structure and the synchronization mechanism employed (i.e. rendezvous or mutual exclusion instances) while taking into account resource conflict avoidance constraints are quite unique (Wójcik 2007).

In that context our main contribution is to propose a new modeling framework enabling to evaluate the cyclic steady state of a given system of concurrent cyclic processes (SCCP). The following questions are of main interest: Does the assumed system behavior (e.g. cyclic steady states) can be achieved under the given system's structure constraints? and if yes: Whether the available states are mutually *reachable from each other*?

So, the chapter's objective is to provide the sufficient conditions for cyclic steady states generation in a system composed of concurrently flowing cyclic processes interacting between each other on the base of a mutual exclusion protocol. This objective regards of quite large class of digital and/or logistics networks that share common properties even though they have huge intrinsic differences. The most important property concerns of different sub-networks infrastructure enabling to schedule multimodal processes executed through connected parts of different local networks (Bocewicz and Banaszak 2013; Bocewicz et al. 2011a). The passenger's itinerary including different metro lines encompass a plan of multimodal process execution within a considered metro network just regards of multimodal processes scheduling problem.

In other words, this study aims to present a declarative approach to define model of reachability problem that can be used further to assist decision-makers in generation, analyzing and evaluating of cyclic steady states reachable in a given SCCP structure. An illustrative model of the constraint satisfaction problem implemented in Oz Mozart language is discussed from multimodal processes perspective.

The rest of the chapter is organized as follows: The Sect. 2 introduces to the systems of concurrently flowing local cyclic and multimodal processes, provides notation used and states a problem. In the Sect. 3 some feasible states space generation issues are discussed and two methods aimed at cyclic steady states refinement are provided. The two cases illustrating both methods implementation are discussed in the Sect. 4. Conclusions are presented in the Sect. 5.

2 Systems of Concurrent Cyclic Processes

2.1 Declarative Modeling

Automated Guided Vehicles System (AGVS) with distinguished vehicles, pick-up/ delivery points (PDPs), and transportations routes is shown in Fig. 1. This kind of system can be modeled in terms of SCCPs, wherein the cyclic multimodal processes (representing the transportations routes) are executed along the parts of cyclic local processes (represented by vehicles itinerary) which are interconnected each other through AGVS common resources representing PDPs. Figure 2 presents the SCCP's model from Fig. 1.

Six **local cyclic processes** (vehicles) are considered $P_1, P_2, P_3, P_4, P_5, P_6$. The processes follow the **routes** composed of transportation sectors. Besides of local processes two **multimodal processes** (i.e. processes executed along the routes composed of parts of local processes): mP_1, mP_2, are considered as well. Processes P_4, and P_5 contain two sub-processes (streams P_j^i) $P_4 = \{P_4^1, P_4^2\}$, $P_5 = \{P_5^1, P_5^2\}$, respectively, i.e. processes (vehicles) moving along the same route. Rest of local processes contain unique streams: $P_1 = \{P_1^1\}$, $P_2 = \{P_2^1\}$,

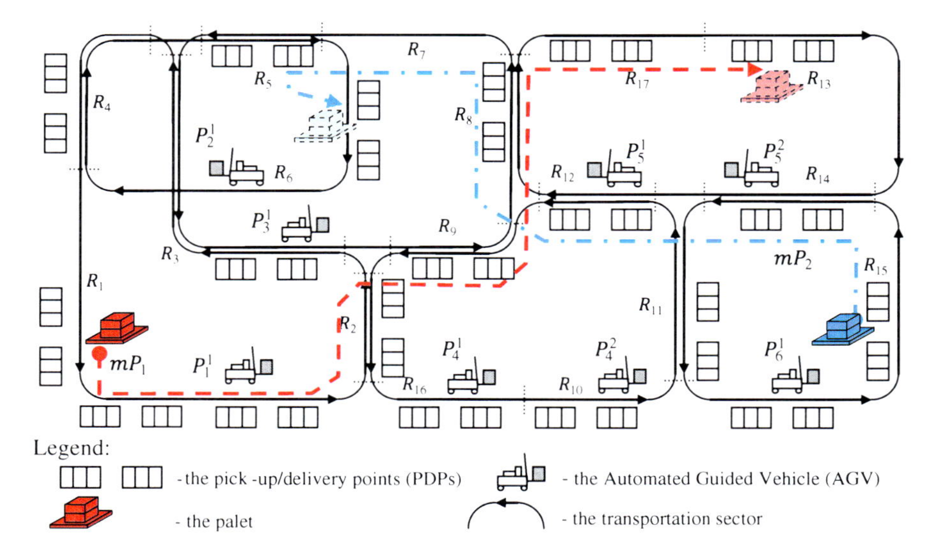

Fig. 1 An example of AGV system

$P_3 = \{P_3^1\}$, $P_6 = \{P_6^1\}$. Processes can interact each other through common shared resources, i.e. transportation sectors. Routes of local processes considered are as follows:

$$p_1^1 = (R_1, R_2, R_3, R_4),\ p_2^1 = (R_4, R_5, R_6),$$
$$p_3^1 = (R_3, R_9, R_8, R_7, R_5),\ p_4^1 = p_4^2 = (R_2, R_{16}, R_{10}, R_{11}, R_{12}, R_9),$$
$$p_5^1 = p_5^2 = (R_8, R_{17}, R_{13}, R_{14}, R_{12}),\ p_6^1 = (R_{11}, R_{15}, R_{14}),$$

where: $R_2 - R_5$, $R_8 - R_{14}$, R_{16}, R_{17} are shared resources, since each one is used by at least two streams (i.e. R_{13} and R_{17} are used by p_5^1, p_5^2), and R_1, R_6, R_7, R_{15}, are non-shared ones because are exclusively used by only one stream. Note that streams p_4^1, p_4^2, belonging to P_4 and p_5^1, p_5^2, belonging to P_5 follow the same route (these streams correspond to vehicles moving along the same route).

Consider two cyclic multimodal processes mP_1, mP_2, following the routes mp_1, mp_2, respectively (see Fig. 2):

$$mp_1 = (R_1, R_2, R_3, R_9, R_8, R_{17}, R_{13}),$$
$$mp_2 = (R_{15}, R_{14}, R_{12}, R_8, R_7, R_5, R_6).$$

Let as assume that multimodal processes do not contain sub-processes, i.e. that each multimodal process consists of a unique stream.

The class of the SCCP considered follows the constraints stated below (Bocewicz and Banaszak 2013):

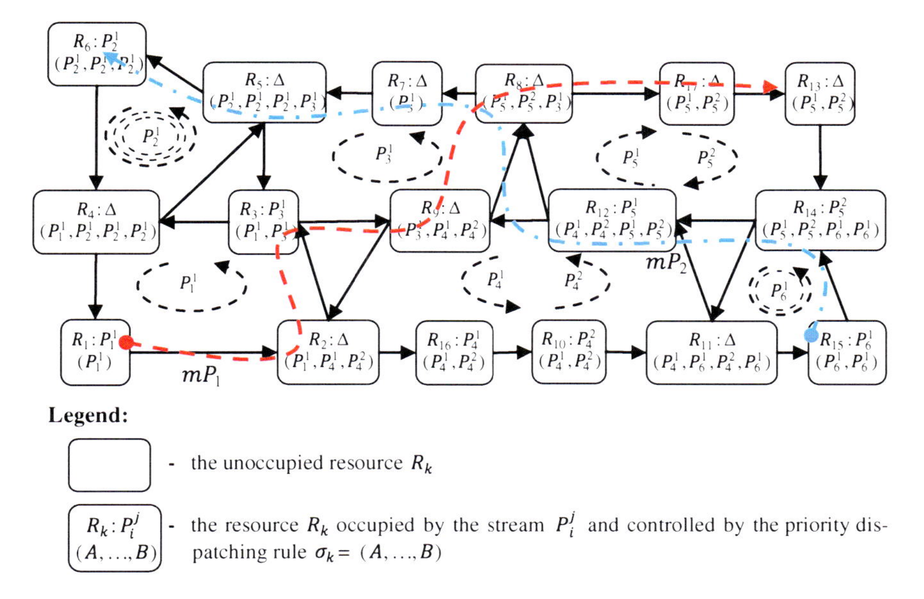

Fig. 2 An example of the FMS: SCCP model of AGVS

- the new operation may start on a resource only if the current operation has been completed and the resource has been released,
- the local processes share the common resources in the mutual exclusion mode, the local process operation can be suspended only if designed resource is occupied, the suspended local processes cannot be released, local processes are non-preempted, i.e. the resource may not be taken of a process till it is using it,
- the multimodal processes encompassing pallets flow conveyed by AGVs follow the local transportation routes, different multimodal processes can be executed simultaneously along the same local process,
- the local and multimodal processes execute cyclically with periods Tc and Tm, respectively; resources occur uniquely in each transportation route,
- in a cyclic steady state, each the ith stream has to pass its local route the same number of times $\Xi \cdot \psi_i$, the factors Ξ, ψ_i, are defined below.

A resource conflict (caused by mutual exclusion protocol usage) is resolved with help of a priority dispatching rule (Bocewicz et al. 2009a) determining an order in which streams make their access to common shared resources (for instance, in case of the resource R_9, $\sigma_9 = \left(P_3^1, P_4^1, P_4^2\right)$—the priority dispatching rule determines the order in which streams can access to the shared resource R_9, i.e. at first to the stream P_3^1, then to the stream P_4^1, next to P_4^2 and once again to P_3^1, and so on). The stream P_i^k occurs the same number of times in each dispatching rule associated to resources appearing in its route. So, the SCCP shown in Fig. 2 is specified by the following set of dispatching rules $\Theta = \{\sigma_1, \ldots, \sigma_{17}\}$, as well as

$f_1(P_1^1) = f_2(P_1^1) = f_3(P_1^1) = f_4(P_1^1) = 1, f_4(P_2^1) = f_5(P_2^1) = f_6(P_2^1) = 3$, etc. where $f_i(P_j)$—a number the jth process occurs in the ith priority dispatching rule. That means the stream P_2^1 repeats 3 times (that is guaranteed due $\sigma_6 = (P_2^1, P_2^1, P_2^1))$ while P_1^1 only ones during the same period $(\sigma_1 = (P_1^1))$.

It means the priority rules determine frequencies of mutual appearance of local processes sharing the same resource. In general case, the set of dispatching rules Θimplies the sequence of relative frequencies of local processes mutual executions, and denoted by $\Psi = (\psi_1, \psi_2, \ldots, \psi_n)$, where: $\psi_i \in \mathbb{N}$,

$$\psi_i = \| \{b | crd_b \sigma_a = P_i^1; b \in \{1, \ldots, lp(a)\}\} \| \quad \forall i \in \{1, \ldots, n\}, \forall \sigma_a \in \Theta_i \tag{1}$$

where: Θ_i—the set of dispatching rules associated to resources occurring in the route followed by P_i, $crd_b \sigma_a$,—the bth entry of the sequence σ_a, n,—a number of processes, $lp(a)$—the length of σ_a.

So, the SCCP shown in Fig. 2 is specified by the sequence: $\Psi = (1, 3, 1, 1, 1, 2)$. That means one execution of local processes P_1, P_3, P_4 , P_5, falls on three executions of process P_2, and two executions of P_6.

Since the sequence Ψ of relative frequencies of local processes mutual executions does not necessary encompass cyclic steady state of a SCCP, hence a new parameter describing the number of x occurrences within a cyclic steady state, denoted by $\Xi \in \mathbb{N}$, is introduced. For the considered SCCP, the value $\Xi = 2$, means that two executions of the sequence $\Psi = (1, 3, 1, 1, 1, 2)$, i.e., two executions of local processes P_1, P_3, P_4, P_5 fall on six executions of the process P_2, and four executions of P_6. In a similar way the mutual frequency $m\Psi$ of multimodal processes mP_i (for example in case of the execution of multimodal processes—see SCCP from Fig. 2—$m\Psi = (1, 1)$—means that one execution of the process mP_1 falls on one execution of mP_2) as well as $m\Xi$ determining a number of $m\psi$ execution in a cycle can be also defined.

In general case, the following notations are used:

- a sequence $p_i^k = \left(p_{i,1}^k, p_{i,2}^k, \ldots, p_{i,lr(i)}^k\right)$ specifies **the route of the local process's stream P_i^k** (kth stream of the ith local process P_i), and its components define the resources used in course of operations execution, where: $p_{i,j}^k \in R$ (the set of resources: $R = \{R_1, R_2, \ldots, R_m\}$)—denotes the resource used by the kth stream of the ith local process in the jth operation; in the rest of the chapter **the jth operation executed on resource $p_{i,j}^k$ in the stream P_i^k** will be denoted by $o_{i,j}^k$, $lr(i)$—denotes a length of cyclic process route (each stream's route p_i^k of P_i has the same length).
- $x_{i,j,q}^k(l) \in \mathbb{N}$—the moment the operation $o_{i,j}^k$ starts its qth execution in the lth cycle of the stream P_i^k.

Table 1 Local operation times of SCCP's (from Fig. 2)

Streams	i, k	$t^k_{i,1}$	$t^k_{i,2}$	$t^k_{i,3}$	$t^k_{i,4}$	$t^k_{i,5}$	$t^k_{i,6}$
P^1_1	1, 1	2	2	2	2	–	–
P^1_2	2, 1	2	2	2	–	–	–
P^1_3	3, 1	3	3	3	3	–	–
P^1_4	4, 1	1	1	1	1	1	1
P^2_4	4, 2	1	1	1	1	1	1
P^1_5	5, 1	2	2	2	2	2	–
P^2_5	5, 2	1	1	1	1	1	–
P^1_6	6, 1	1	1	1	–	–	–

- $t^k_i = \left(t^k_{i,1}, t^k_{i,2}, \ldots, t^k_{i,lr(i)}\right)$ specifies **the local process operation times**, where $t^k_{i,j}$ denotes the time of execution of operation $o^k_{i,j}$ (for SCCP from Fig. 2 see Table 1).
- $mp_i = (mpr_j(a_j, b_j), mpr_l(a_l, b_l), \ldots, mpr_h(a_h, b_h))$ specifies **the route of the multimodal process mP_i** where:
- $mpr_j(a, b) = \left(crd_a p^k_j, crd_{a+1} p^k_j, \ldots, crd_b p^k_j\right)$, $crd_i D = d_i$, for $D = (d_1, d_2, \ldots, d_i, \ldots, d_w)$, $\forall a \in \{1, 2, \ldots, lr(i)\}$, $\forall j \in \{1, 2, \ldots, n\}$, $crd_a p_j \in R\cdot$
- The transportation route mp_i is the sequence of sections of local process routes. In the rest of the chapter **the jth operation executed in the process mP_i** will be denoted by $mo_{i,j}$.
- $mx_{i,j,k}(l) \in \mathbb{N}-$ the moment the operation $mo_{i,j}$ starts its kth execution in the lth cycle.
- $\Theta = \{\sigma_1, \sigma_2, \ldots, \sigma_m\}$ is the set of **the priority dispatching rules**, where $\sigma_i = (s_{i,1}, \ldots, s_{i,lp(i)})$ is the sequence components of which determine an order in which the processes can be executed on the resource R_i, $s_{i,j} \in P$ (the set of local process streams).

Using the above notation a SCCP can be defined as a tuple:

$$SC = ((R, SL), SM) \tag{2}$$

where:

$R = \{R_1, R_2, \ldots, R_m\}$	the set of resources,
m	the number of resources,
$SL = (ST_L, BE_L)$	the local processes structure, i.e
$ST_L = (U, T)$	the variables describing layout of local processes,
$U = \left\{p^1_1, \ldots, p^{ls(1)}_1, \ldots, p^1_n, \ldots, p^{ls(n)}_n\right\}$	the set of local process routes,
$ls(i)$	the number of streams belonging to the process P_i,
n	a number of local processes,

$T = \left\{t_1^1, \ldots, t_1^{ls(1)}, \ldots, t_n^1, \ldots, t_n^{ls(n)}\right\}$ the set of sequences of operation times in local processes

$BE_L = (\Theta, \Psi, \Xi)$ the variables describing the local processes behavior,

$\Theta = \{\sigma_1, \sigma_2, \ldots, \sigma_m\}$ the set of dispatching priority rules,

$\Psi = (\psi_1, \psi_2, \ldots, \psi_n)$ the sequence of relative frequencies of local processes mutual executions,

Ξ the number of Ψ occurrences within a cyclic steady state

$SM = (ST_M, BE_M)$ the multimodal processes structure, i.e

$ST_M = (M, T)$ the variables describing layout of multimodal processes level,

$M = \{mp_1, \ldots, mp_w\}$ the set of multimodal process routes,

w the number of multimodal processes mP_i,

$mT = \{mt_1, \ldots, mt_w\}$ the set of sequences of operation times in multimodal processes,

$BE_M = (m\Psi, m\Xi)$ the variables describing the multimodal processes behavior,

$\{m\Psi = (\psi_1, \psi_2, \ldots, \psi_w)\}$ the sequence of relative frequencies of multimodal processes mutual executions,

$m\Xi$ the number of $m\Psi$ occurrences within a cyclic steady state.

The SCCP model (2) can be seen as a multi-level one, see Fig. 3, i.e. composed of the "*R* level" (resources), the "*SL* level" (local cyclic processes), and the "SM^1 level" (multimodal cyclic processes) as well as the "SM^i level" (the *i*th meta-multimodal process). *SL* level determines a structure of local processes transportation routes *U* as well as parameters Θ, Ψ, Ξ specifying required system's behavior. In turn, the SM^1 level enables to consider the multi-modal processes, as well as meta-multimodal processes (SM^2 level) composed of multimodal processes from SM^1 level.

So, in general case $SC = ((R, SL), SM)$ model can be seen as composed of *i* levels:

$$SC^i = (((((R, SL), SM^1), SM^2), \ldots), SM^i) \tag{3}$$

Note, that cyclic behavior of the SC^{i-1} implies the periodic behavior of SC^i, too.

2.2 Problem Formulation

Consider a SCCP specified by the given set *R* of resources, dispatching rules Θ, local and multimodal processes routes *U*, *M* and initial processes allocation. The main question concerns of SCCP periodicity, i.e. does the cyclic execution of local processes exist? In case when they are periodic the another question can be stated:

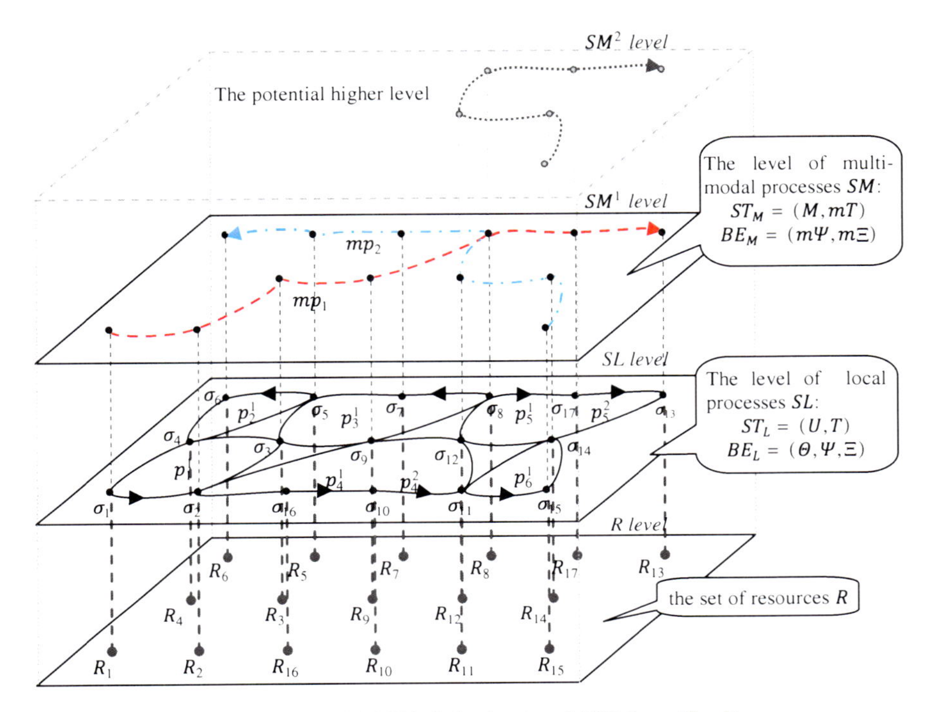

Fig. 3 The multilayered model of SCCP's behavior (see SCCP from Fig. 2)

What is the period Tc? The other questions regard of multimodal processes cyclic execution.

Response to above questions require answers to more detailed questions, for instance: What are admissible initial processes allocations (i.e. the possible AGVs dockings)? What are dispatching rules Θ guaranteeing a given SCCP periodicity (in local and multimodal sense) while preserving assumed frequency (Ψ, $m\Psi$) of processes execution within a global period (local Tc and multimodal Tm)? In general case, however, besides of above mentioned so called forward problem formulations the backward ones can be considered as well. For instance: Does there exist the SCCP's structure of local (ST_L) and/or multimodal layer (ST_M) such that an assumed steady cyclic state can be achieved?

The problems stated above have been studied in (Bocewicz and Banaszak 2013; Bocewicz et al. 2009b, 2011a, b).

A new problem regarding possible switching among cyclic steady states can be seen as their obvious consequence. In that context, the newly arising questions are: Is it possible to reschedule cyclic schedules as to "jump" from one cyclic steady state to another? Is it possible to "jump" directly or indirectly? What are the control rules allowing one to do it? These kind of questions are of crucial importance for manufacturing and transportation systems aimed at short run production and/or passengers itinerary (e.g. in a sub-way network) planning.

3 Agvs Schedulability

3.1 States Space Generation

Consider the following SCCPs state definition describing both the local and multimodal processes allocation:

$$S^k = \left(Sl^r, MA^k\right) \tag{4}$$

where:

- Sl^r—is the state of local processes, corresponding to the kth state of multimodal processes,

$$Sl^r = (A^r, Z^r, Q^r) \tag{5}$$

- where: $A^r = \left(a_1^r, a_2^r, \ldots, a_m^r\right)$—the processes allocation in the rth state, $a_i^r \in P \cup \{\Delta\}$, $a_i = P_j^k$—the ith resource R_i is occupied by the local stream P_j^k, and $a_i^r = \Delta$—the ith resource R_i is unoccupied.
- $Z^r = \left(z_1^r, z_2^r, \ldots, z_m^r\right)$—the sequence of semaphores corresponding to the rth state, $z_i^r \in P$—means the name of the stream (specified in the ith dispatching rule σ_i, allocated to the ith resource) allowed to occupy the ith resource; for instance $z_i^r = P_j^k$ means that at the moment stream P_j^k is allowed to occupy the ith resource.
- $Q^r = \left(q_1^r, q_2^r, \ldots, q_m^r\right)$—the sequence of semaphore indices, corresponding to the rth state, q_i^r determines the position of the semaphore z_i^r in the priority dispatching rule σ_i, $z_i^r = crd_{\left(q_i^r\right)}\sigma_i$, $q_i^r \in \mathbb{N}$. For instance $q_2^r = 2$ and $z_2^r = P_1^2$, that means the semaphore $z_2^r = P_1^2$ takes the 2nd position in the priority dispatching rule σ_2.
- MA^k—the sequence of multimodal processes allocation: $MA^k = \left(mA_1^k, \ldots, mA_u^k\right)$, mA_i^k—allocation of the process mP_i, i.e.:

$$mA_i^k = \left(ma_{i,1}^k, ma_{i,2}^k, \ldots, ma_{i,m}^k\right) \tag{6}$$

where: m—is a number of resources R, $ma_{i,j}^k \in \{mP_i, \Delta\}$, $ma_{i,j}^k = mP_i$ means, the jth resource R_j is occupied by the ith multimodal process P_i, and $ma_{i,j}^k = \Delta$—the ith resource R_j is released by the ith multimodal process P_i.
In that context, **the state S^k is feasible** (Bocewicz and Banaszak 2013) when:

- the semaphores of occupied resources indicate the streams allocated to those resources,
- each local/multimodal stream is allotted to a unique resource due to relevant local/multimodal process route.

The introduced concept of the kth state $\boldsymbol{S^k}$ enables to create a space $\mathbb{S}$ of feasible states. For the purpose of illustration let us consider the state space of the SCCP composed of 6 resources, and 3 local cyclic processes supporting one multimodal process (see Fig. 4). The observed behavior is two folded, i.e. the levels of local SL, and multimodal processes SM can be distinguished. In case of local processes level, states Sl^i are noted by "O", and at the multimodal level the relevant states $S^i \in \mathbb{S}$ are noted by "O"). The states $Sl^j \in \mathbb{S}l$ can be considered as a part of associated states $S^i \in \mathbb{S}$, i.e. the states being elevation of relevant states S^i.

Transitions linking feasible states S^k, $S^l \in \mathbb{S}$, while following non-preemption and mutual exclusion constraints are denoted by $S^k \rightarrow S^l$, and encompass the next state function δ: $S^l = \delta\left(S^k\right)$, definition of which (Bocewicz et al. 2011a) leads to the following property:

Property 1

Each $S^i \in \mathbb{S}$ *can be preceded by some subset of states* $\mathbb{SP}^i$, $\mathbb{SP}^i \subset \mathbb{S}$, *(also* $\mathbb{SP}^i = \emptyset$*), i.e.* $\forall S^k \in \mathbb{SP}^i$, $S^i = \delta\left(S^k\right)$, *but can result only in a unique state* $S^j \in \mathbb{S}$, *i.e., there exists at most one* $S^j \in \mathbb{S}$, $S^j = \delta(S^i)$.

The deadlock state $S^* \in \mathbb{S}$ resulting in the SCCP blockade is free from any descendent state. In that context two kinds of steady state behaviors can be considered: a **cyclic steady state** and a **deadlock state**.

The set $mSc^* = \left\{S^{k_1}, S^{k_2}, S^{k_3}, \ldots, S^{k_v}\right\}$, $mSc^* \subset \mathbb{S}$, is called **a reachability state space of multimodal processes** generated by an initial state $S^{k_1} \in \mathbb{S}$, if the following condition holds:

$$S^{k_1} \overset{i-1}{\rightarrow} S^{k_i} \overset{v-i-1}{\rightarrow} S^{k_v} \rightarrow S^{k_i} \tag{7}$$

where: $S^a \overset{i}{\rightarrow} S^b$—the transition defined in (Bocewicz and Banaszak 2013), $S^{k_1} \overset{i}{\rightarrow} S^{k_{i+1}} \equiv S^{k_1} \rightarrow S^{k_2} \rightarrow S^{k_3} \rightarrow \cdots \rightarrow S^{k_{i+1}}$

The set $mSc = \left\{S^{k_i}, S^{k_{i+1}}, \ldots, S^{k_v}\right\}$, $mSc \subseteq mSc^*$, is called **a cyclic steady state of multimodal processes** (i.e., a cyclic steady state of $SCCP$) with the period $Tm = \|mSc\|$, $Tm > 1$, In other words a cyclic steady state contains such a set of states in which starting from any distinguished state it is possible to reach the rest of states and finally reach this distinguished state again:

$$\forall_{S^k \in mSc}\left(S^k \overset{Tm-1}{\longrightarrow} S^k\right) \tag{8}$$

The cyclic steady state Sc specified by the period Tc of local processes execution is defined in the similar way. Graphically the cyclic steady states Sc and mSc are described by cyclic and spiral digraphs, respectively, see Fig. 4.

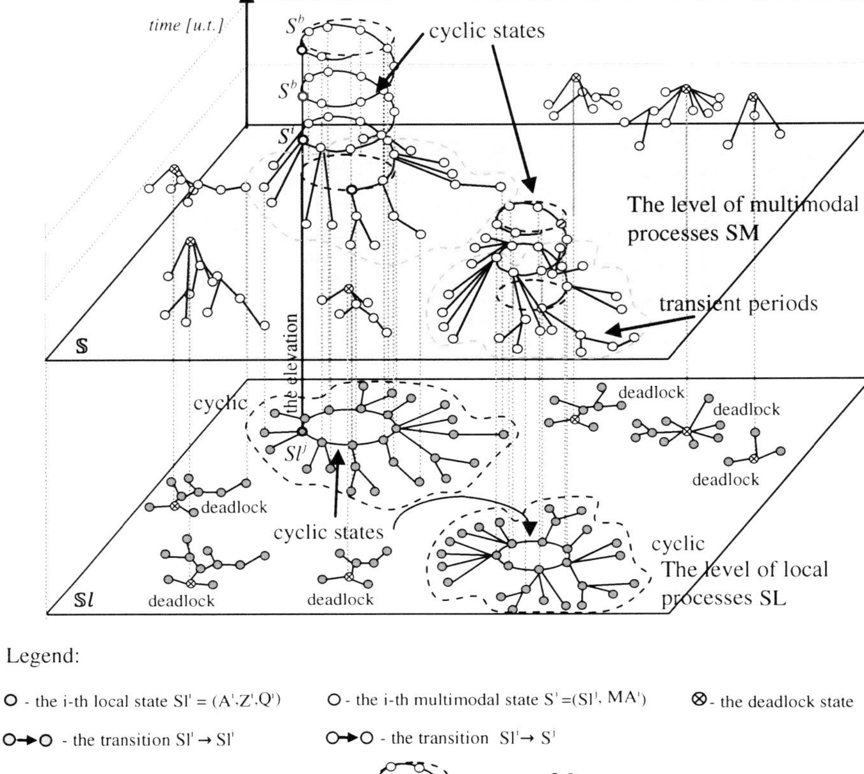

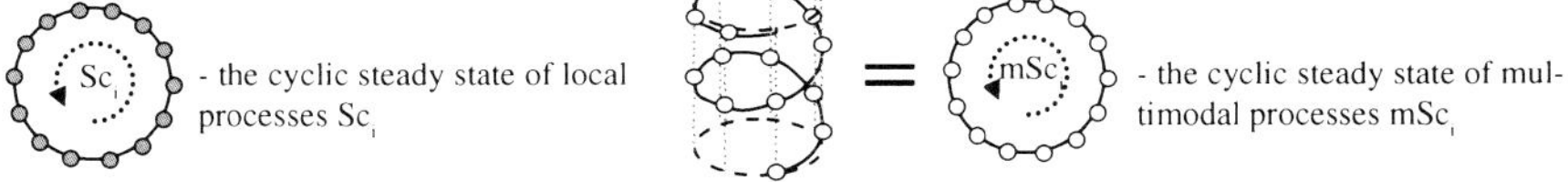

Fig. 4 The space of feasible states encompassing the SCCP's behavior (see SCCP from Fig. 2)

Moreover, since an initial state $S^{k_1}\mathbb{S}$ either lead to *mSc* or to a deadlock state S^*, i.e. $S^{k_1} \xrightarrow{i-1} S^{k_i} \xrightarrow{v-i-1} S^{k_v}*$, hence multimodal processes also can reach a **deadlock state**, denoted by "$\otimes$" in Fig. 4.

In that context, our question regarding periodicity of SCCP results in the question whether there exists an initial state S^0 generating the cyclic steady state *mSc*. It means, that searching for a cyclic steady state *mSc* in a given SCCP can be seen as a reachability problem where for an assumed initial state S^0 (i.e. determining local and multimodal processes allocations) the state S^k, such that following transitions $S^0 \xrightarrow{i} S^k \xrightarrow{Tm-1} S^k$ holds, is sought.

The reachability problem stated for a cyclic steady states space encompassing multimodal processes behavior can be observed in many real life cases, i.e. the trot and gallop gaits, sub-way schedules for pick and off-pick hours and so on.

Assuming the considered multimodal processes behavior consist of a set of potentially reachable cyclic steady states the conditions guaranteeing switching among them play a primary role in smooth cyclic schedules rescheduling. Therefore, the rest of the chapter is devoted to the following reachability problem of multimodal processes cyclic steady states space:

Given is the SC specified by (2), *i.e.,* R, $SL = (ST_L, BE_L, SE_L), M = (ST_M, BE_M, SE_M)$. *Two cyclic steady states* mSc_1 *and* mSc_2 *of the SC (encompassing cyclic steady states of local and multimodal processes) are known. Is the cyclic steady state* mSc_2 *reachable from the* mSc_1*?*

So, the question we are facing with is: Is it possible to switch directly or indirectly from one cyclic steady state of multimodal processes to an assumed another one? For instance, let us consider cyclic steady states mSc_1 and mSc_2 from Fig. 4.

Searching for direct switching between mSc_1 and mSc_2 assumes the state $S^x \in \mathbb{S}$ belonging to both cyclic steady states has to exist, that means due to (8) the following transitions have to fulfill:

- $\forall_{S^k \in mSc_1} \left(S^k \xrightarrow{i} S^x \xrightarrow{Tm_1 - 1 - i} S^k \right)$ for the cyclic steady state mSc_1,
- $\forall_{S^l \in mSc_2} \left(S^l \xrightarrow{i} S^x \xrightarrow{Tm_2 - 1 - i} S^l \right)$ for the cyclic steady state mSc_2.

That is impossible because S^x cannot has to have two descendents. What is impossible at the $\mathbb{S}l$ and $\mathbb{S}$ levels can be possible, however, at the $\mathbb{A}$ level, see Fig. 5. At the level $\mathbb{A}$ there are allocations A^i possessing more than one descendent. Such situation corresponds to an allocation belonging to the several states Sl^i. For instance, A^2 belongs to $Sl^3 = (A^2, Z^3, Q^3)$ and $Sl^4 = (A^2, Z^4, Q^4)$, simultaneously. The same regards of A^1 belonging to Sl^1 and Sl^2. The states of local processes specified by the common allocation are different because the semaphores and indices are different (Z^3, Q^3 and Z^4, Q^4).

This observation can be employed in the course of searching for states enabling switching between assumed cyclic steady states. In the case considered the direct switching between mSc_2 and mSc_1 is allowed for the states possessing so called shared allocation, e.g. A^2, see $Sl^4 = (A^2, Z^4, Q^4)$. So, replacing at this allocation Z^4, Q^4 by Z^3, Q^3 results in switching from Sl^4 to Sl^3. That implies the possible direct switching between cyclic multimodal processes exist.

Such rules of semaphores and indices changes can be treated as relevant control rules. The similar control rules may involve the priority dispatching rules Θ changing. In both cases changes do not require any allocations change, and then do not lead to the system stoppage.

Consequently, the following properties can be stated:

Property 2

The cyclic steady state $mSc_1 \subseteq mSc_1^*$ *is reachable from the cyclic steady state* $mSc_2 \subseteq mSc_2^*$, *(denoted by* $mSc_2 \rightarrow mSc_1$*), in the SC, only in the case both states* $S^a \in mSc_1^*$ *and* $S^b \subseteq mSc_2$ *possess the same allocation* A^x.

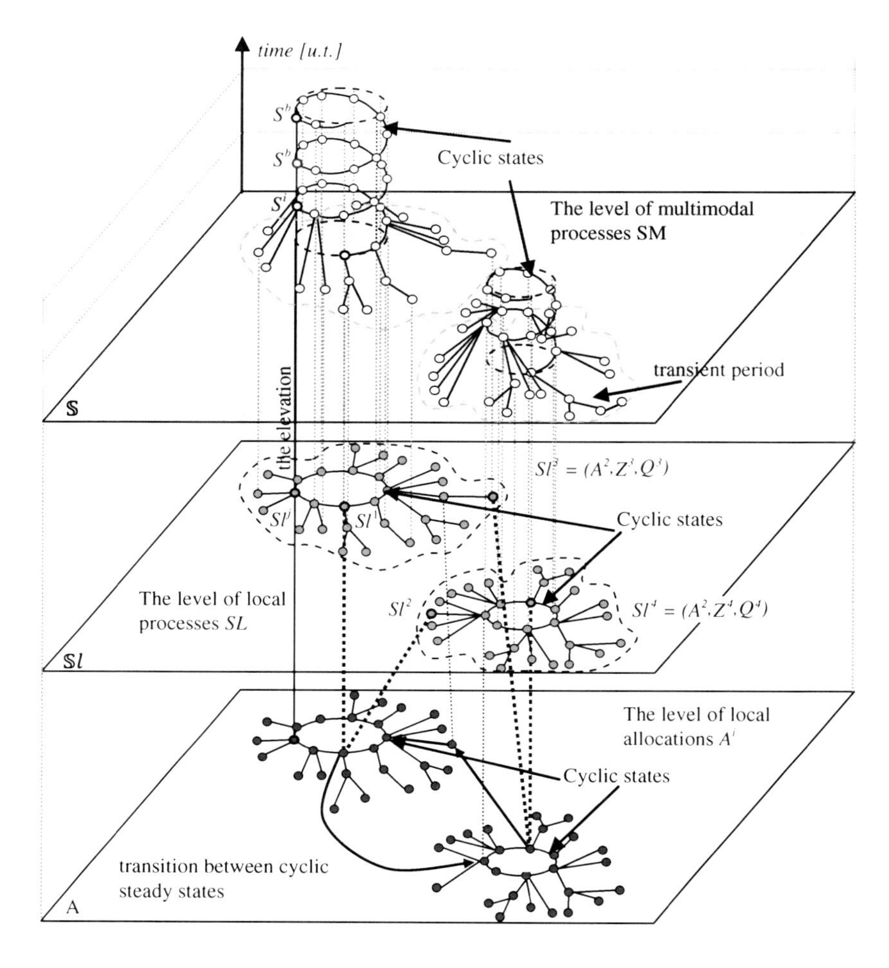

Legend:

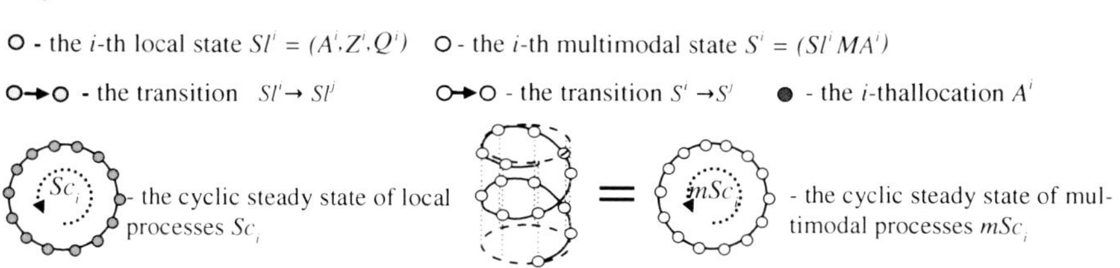

Fig. 5 The space of feasible states encompassing the SCCP's behavior (see SCCP from Fig. 2)

Property 3

Two cyclic steady states $mSc_1 \subseteq mSc_1^*$ *and* $mSc_2 \subseteq mSc_2^*$ *from the SC are mutually reachable, (denoted as* $mSc_2 \leftrightarrow mSc_1$*) only if,* $mSc_1 \rightarrow mSc_2$ *and* $mSc_2 \rightarrow mSc_1$ *hold.*

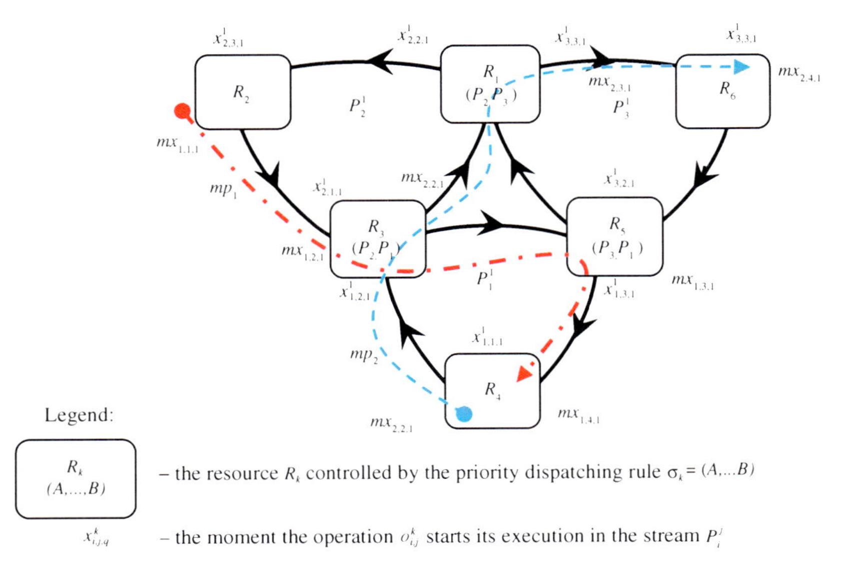

Fig. 6 The SCCP with dispatching rules: $\sigma_1 = (P_2, P_3), \sigma_3 = (P_2, P_1), \sigma_5 = (P_3, P_1)$, and $\Psi = (1, 1, 1)$, $\Xi = 1$

So, the reachability problem of the cyclic steady states space, e.g. regarding of switching between two states $mSc_1 \subseteq mSc_1^*$ and $mSc_2 \subseteq mSc_2^*$, concludes in the question: Does there exist two states $S^a \in mSc_1$ and $S^b \subseteq mSc_2^*\, S^a \in mSc_1^*$ (and $S^b \subseteq mSc_2$) sharing the same allocation A^x of local cyclic processes?

Note that refining the cyclic steady states space from a given feasible states space is quite easy. However, the problem of feasible states space generation is NP-hard. The majority of states are either deadlocks or just leading to the deadlock states. So, in order to avoid generation of the whole feasible states space let us focus on an alternative approach aimed at dedicated generation of **cyclic steady states.**

3.2 *Cyclic Steady States Space Generation*

Since parameters describing the SCCP are usually discrete, and linking them relations can be seen as constraints, hence related to them cyclic scheduling problems can be presented in the form of the Constraint Satisfaction Problem (*CSP*) (Bocewicz et al. 2009a; Bocewicz and Banaszak 2013; Schulte et al. 1998). More formally, *CSP* is a framework for solving combinatorial problems specified by pairs: (a set of variables and associated domains, a set of constraints restricting the possible combinations of the variable values). So, in case of SC (2) the CSP is defined as follows:

$$CS(SC) = ((\{X, Tc, mX, Tm\}, \{D_X, D_{Tc}, D_{mX}, D_{Tm}\}), C) \tag{9}$$

where:

X, Tc, mX, Tm—the decision variables, where Tc and Tm are local and multimodal periodicities;$X = \left\{X_1^1, \ldots, X_1^{ls(1)}, \ldots, X_n^1, \ldots, X_n^{ls(n)}\right\}$—the set of sequences of X_i^k, and $X_i^k = \left(x_{i,1,1}^k, \ldots, x_{i,lr(i),1}^k, \ldots, x_{i,1,\Xi}^k, \ldots, x_{i,lr(i),\Xi}^k\right)$, $x_{i,j,q}^k$—the moment of operation $o_{i,j}^k$ (local process) beginning in the first cycle, $x_{i,j,q}^k$ and $x_{i,j,q}^k(l)$ are linked by: $x_{i,j,q}^k(l) = x_{i,j,q}^k + l \cdot Tc$, $l \in \mathbb{Z}$, $Tc = x_{i,j,q}^k(l+1) - x_{i,j,q}^k(l)$,

Analogously $mX = \{mX_1, mX_2 \ldots, mX_w\}$—is the set of sequences of mX_i, and $mX_i = (mx_{i,1,1}, \ldots, mx_{i,lm(i),1}, \ldots, mx_{i,1,\Xi}, \ldots, mx_{i,lm(i),\Xi})$, $mx_{i,j,k}$—the moment of operation $mo_{i,j}$ (of multimodal process) beginning in the first cycle $mx_{i,j,k}(l) = mx_{i,j,k} + l \cdot Tm$, and $Tm = mx_{i,j,k}(l+1) - mx_{i,j,k}(l)$,

the following domains of decision variables are considered:

D_X, D_{mX}—the family of sets of admissible entry values X_i, $x_{i,j,q}^k \in \mathbb{Z}$, and mX_i, $mx_{i,j,k} \in \mathbb{Z}$,$D_{Tc}, D_{Tm}$—the domains of the variable $Tc \in \mathbb{N}$ and $Tm \in \mathbb{N}$.

C—constraints are specified by both:

- $ep_{i,j,q}^k(ST_L, BE_L)$—the set of constraints (equations) linking ST_L (local processes structure) and BE_L (local processes behavior). Each $ep_{i,j,q}^k(ST_L, BE_L)$ describes, the time relation [according above presented conditions (Bocewicz and Banaszak 2013)] between the moments of operations beginning for its qth execution: $i = 1, .., n; j = 1, \ldots, lr(i); k = 1, .., ls(i); q = 1, \ldots, \Xi$.
- $eq_{i,j,k}(ST_M, BE_M)$—the set of constraints (equations) linking ST_M (multimodal processes structure) and BE_M (multimodal processes behavior). Each $eq_{i,j,k}(ST_M, BE_M)$ describes, the time relation between the moments of multimodal operations beginning for its kth execution: $i = 1, .., w$; $j = 1, \ldots, lm(i)$; $k = 1, \ldots, m$.

Solution to the problem (9) provides a set of X, mX, sequences values of which guarantee the required cyclic behavior of a SCCP while following the set of constraints C. The conditions C sufficient for SCCP cyclic behavior (resulting in collision-free and deadlock-free (Lawley et al. 1998) processes execution) are formulated using the operator *max* and are introduced in (Bocewicz and Banaszak 2013; Bocewicz et al. 2011a).

3.2.1 Constraints of Local Processes

In order to explain the way the constraints of local processes are designed let us consider an example of SCCP shown in Fig. 6. The operation $o_{1,3}^1$ (executed by P_1^1

on the resource R_5) can be started (i.e., beginning its first execution; $q = 1$) only if the preceding operation $o^1_{1,2}$ (executed by P^1_1 on R_3) has been completed $\left(x^1_{1,2,1} + t^1_{1,2}\right)$ and the resource R_5 has been released, i.e. if the stream P^1_3 occupying the resource R_5 starts its subsequent operation at $x^1_{3,3,1} + 1$. So, the relation considered $ep^1_{1,3,1}(ST_L, BE_L)$ can be specified by the following formulae:

$$x^1_{1,3,1} = \max\left\{\left(x^1_{3,3,1} + 1\right); \left(x^1_{1,2,1} + t^1_{1,2}\right)\right\} \tag{10}$$

where: $x^k_{i,j,q}$—the moment of operation $o^k_{i,j}$ (local process) beginning in qth execution.

Besides of (10) the Table 2 contains the rest of constraints describing the local processes of SCCP from Fig. 6. For all constraints the following principle holds: the moment of the operation $o^k_{i,j}$ beginning states for a maximum of the completion time of the operation $o^k_{i,j-1}$ preceding $o^k_{i,j}$, and the release time of the resource $p^k_{i,j}$ awaiting for the $o^k_{i,j}$ execution.

$$\begin{aligned}
&ep^k_{i,j,q}(ST_L, BE_L):\\
&\textit{moment of operation } o^k_{i,j} \textit{ begining (for } q\textit{-th execution)}\\
&\quad = \max\left\{\textit{moment of } p^k_{i,j} \textit{ release, moment of operation } o^k_{i,j-1} \textit{ completion}\right\}\\
&\qquad i = 1, .., n; j = 1, \ldots, lr(i); k = 1, .., ls(i); q = 1, \ldots, \Xi
\end{aligned} \tag{11}$$

Therefore, following the constraints (11) guarantees deadlock-free execution of cyclic processes at the SL level (see Fig. 3). The number n_{ep} of constraints $ep^k_{i,j,q}$ follows from the number of processes n, streams $ls(i)$, operations $lr(i)$ as well as $\Xi \cdot n_{ep} = \Xi \cdot \sum_{i=1}^{n}[lr(i) \cdot ls(i)]$.

3.2.2 Constraints of Multimodal Processes

The constraints determining local cyclic processes execution has been already discussed in deep (Bocewicz et al. 2009b, 2011a, b). The similar discussion regarding multi processes however has not been done yet. So, for the sake of simplicity let us assume that multimodal processes are collision- and deadlock-free. Multimodal processes may occupy the same resource at a time as well as may use the same local process for their execution simultaneously.

The constraints describing relationship among moments of beginning of successive operations are show on example of mP_1 form Fig. 6. The process considered executes on the set of following resources R_2, R_3, R_5, R_4. The local cyclic process P^1_2 support mP_1 execution between R_2, R_3 and P^1_1 support mP_1 execution between R_3, R_5, R_4. That means the process mP_1 can executes its operations only at

Table 2 Constraints determining the moments $x^k_{i,j,q}$ for the SCCP from Fig. 6

$x^1_{1,1,1} = \max\left\{\left(x^1_{1,3,1} + t^1_{1,3} - Tc\right); \left(x^1_{1,3,1} + 1 - Tc\right)\right\}$
$x^1_{2,2,1} = \max\left\{\left(x^1_{2,1,1} + t^1_{2,1}\right); \left(x^1_{3,1,1} + 1\right)\right\}$
$x^1_{3,1,1} = \max\left\{\left(x^1_{3,3,1} + t^1_{3,3} - Tc\right); \left(x^1_{3,3,1} + 1 - Tc\right)\right\}$
$x^1_{3,3,1} = \max\left\{\left(x^1_{2,3,1} + 1 - Tc\right); \left(x^1_{3,2,1} + t^1_{3,2}\right)\right\}$
$x^1_{1,2,1} = \max\left\{\left(x^1_{1,1,1} + t^1_{1,2}\right); \left(x^1_{2,2,1} + 1\right)\right\}$
$x^1_{2,1,1} = \max\left\{\left(x^1_{2,3,1} + t^1_{2,3} - Tc\right); \left(x^1_{1,3,1} + 1 - Tc\right)\right\}$
$x^1_{2,3,1} = \max\left\{\left(x^1_{2,2,1} + t^1_{2,2}\right); \left(x^1_{2,2,1} + 1\}\right)\right\}$
$x^1_{3,2,1} = \max\left\{\left(x^1_{3,1,1} + t^1_{3,1}\right); \left(x^1_{1,1,1} + 1\right)\right\}$

moments the relevant operations from local processes are performed (i.e. P^1_1 or P^1_2). For instance, the operation $mo_{1,2}$ executed on R_3 depends on P^1_2. The moment $mx_{1,2,1}$ of beginning of the operation $mo_{1,2}$ equals to the moment $\left(x^1_{2,2,1} + a \cdot Tc\right)$ that is the beginning of the operation $o^1_{2,2}$ executed after completion of the preceding operation $mo_{1,2}$ (i.e. operation $mo_{1,1}$ completed at the moment $mx_{1,1,1} + mt_{1,1,1}$). The constraint specifying this relationship has the following form:

$$mx_{1,2,1} = min\left\{\left(x^1_{2,2,1} + a \cdot Tc\right) | a \in \mathbb{Z}, x^1_{2,2,1} + a \cdot Tc \geq mx_{1,1,1} + mt_{1,1,1}\right\}, \quad (12)$$

where: $x^k_{i,j,q}$—the moment of operation $o^k_{i,j}$ beginning in the qth execution of local process, $mx_{i,j,q}$—the moment of operation $mo_{i,j}$ beginning in the qth execution of multimodal process,Tc—the periodicity of local cyclic steady states

Besides of (12) the Table 3 contains the rest of constraints describing multimodal processes of the SCCP from Fig. 6.

For all the constraints the following principle holds: the moment of the operation $mo_{i,j}$ beginning is the earliest moment the operation of the local process can start (of course the one the multimodal process mP_i require), however greater than the moment of operation $mo_{i,j-1}$ beginning. More formally:

$$\begin{aligned} & eq_{i,j,k}(ST_M, BE_M): \\ & \quad \textit{moment of operation } mo_{i,j} \textit{ begining} = \min\{\textit{set of moments of} \\ & \textit{operation } o^k_{i,j-1} \textit{ begining, higher than moment of previous opreation } mo_{i,j-1} \\ & \quad \textit{completion}\}, \ i = 1,..,w; j = 1,\ldots,lm(i); k = 1,\ldots,m\Xi \end{aligned} \quad (13)$$

The number n_{eq} of constraints $eq_{i,j,k}$ equals to: $n_{eq} = m\Xi \cdot \sum_{i=1}^{w} lm(i)$.

Table 3 Constraints determining the moments $mx_{i,j,k}$ for SCCP from Fig. 6

Multimodal process mP_1

$$mx_{1,1,1} = min\left\{\left(x^1_{2,3,1} + a \cdot Tc\right) | a \in \mathbb{Z}, x^1_{2,3,1} + a \cdot Tc \geq mx_{1,4,1} + mt_{1,4,1} - Tm\right\},$$

$$mx_{1,2,1} = min\left\{\left(x^1_{2,2,1} + a \cdot Tc\right) | a \in \mathbb{Z}, x^1_{2,2,1} + a \cdot Tc \geq mx_{1,1,1} + mt_{1,1,1}\right\},$$

$$mx_{1,3,1} = min\left\{\left(x^1_{1,3,1} + a \cdot Tc\right) | a \in \mathbb{Z}, x^1_{1,3,1} + a \cdot Tc \geq mx_{1,2,1} + mt_{1,2,1}\right\},$$

$$mx_{1,4,1} = min\left\{\left(x^1_{1,1,1} + a \cdot Tc\right) | a \in \mathbb{Z}, x^1_{1,1,1} + a \cdot Tc \geq mx_{1,3,1} + mt_{1,3,1}\right\},$$

Multimodal process mP_2

$$mx_{2,1,1} = min\left\{\left(x^1_{1,1,1} + a \cdot Tc\right) | a \in \mathbb{Z}, x^1_{1,1,1} + a \cdot Tc \geq mx_{2,4,1} + mt_{2,4,1} - Tm\right\},$$

$$mx_{2,2,1} = min\left\{\left(x^1_{1,2,1} + a \cdot Tc\right) | a \in \mathbb{Z}, x^1_{1,2,1} + a \cdot Tc \geq mx_{2,1,1} + mt_{1,2,1}\right\},$$

$$mx_{2,3,1} = min\left\{\left(x^1_{2,2,1} + a \cdot Tc\right) | a \in \mathbb{Z}, x^1_{2,2,1} + a \cdot Tc \geq mx_{2,2,1} + mt_{2,2,1}\right\},$$

$$mx_{2,4,1} = min\left\{\left(x^1_{3,1,1} + a \cdot Tc\right) | a \in \mathbb{Z}, x^1_{3,1,1} + a \cdot Tc \geq mx_{2,3,1} + mt_{2,3,1}\right\},$$

$$mod\{Tm, Tc\} = 0$$

4 Periodic Processes Scheduling

Given is an AGVS from Fig. 1 modeled in terms of SCCPs (see Fig. 2). Consider three cases encompassing three different cyclic systems. First of them, due to (2), can be specified as follows:

$$SC^1 = ((R, SL), SM),$$

where

$R = \{R_1, R_2, \ldots, R_{18}\}$—*the set of resources*, $SL = (ST_L, BE_L)$—*the local processes structure*

$$ST_L = (U, T)$$

$$U = \{p^1_1, p^1_2, p^1_3, p^1_4, p^2_4, p^1_5, p^2_5\}$$

$$p^1_1 = (R_1, R_2, R_3, R_4), \quad p^1_2 = (R_4, R_5, R_6), \quad p^1_3 = (R_3, R_9, R_8, R_7, R_5),$$

$$p^1_4 = p^2_4 = (R_2, R_{16}, R_{10}, R_{11}, R_{12}, R_9), \quad p^1_6 = (R_{11}, R_{15}, R_{14}),$$

$$p^1_5 = p^2_5 = (R_8, R_{17}, R_{13}, R_{14}, R_{12}),$$

$T = \{t^1_1, t^1_2, t^1_3, t^1_4, t^2_4, t^1_5, t^2_5\}$, *where values of elements of sequences* t^k_i *are determined in* Table 1

$$BE_L = (\Theta, \Psi, \Xi)$$

$$\Theta = \{\sigma_1, \sigma_2, \ldots, \sigma_{18}\}$$
$$\sigma_1 = (P_1^1), \quad \sigma_2 = (P_1^1, P_4^1, P_4^2), \quad \sigma_3 = (P_1^1, P_3^1),$$
$$\sigma_4 = (P_1^1, P_2^1, P_2^1, P_2^1), \quad \sigma_5 = (P_2^1, P_2^1, P_2^1, P_3^1), \quad \sigma_6 = (P_2^1, P_2^1, P_2^1),$$
$$\sigma_7 = (P_3^1), \quad \sigma_8 = (P_5^1, P_5^2, P_3^1), \quad \sigma_9 = (P_3^1, P_4^1, P_4^2),$$
$$\sigma_{10} = (P_4^1, P_4^2), \quad \sigma_{11} = (P_4^1, P_6^1, P_4^2, P_6^1), \quad \sigma_{12} = (P_4^1, P_4^2, P_5^1, P_5^2),$$
$$\sigma_{13} = (P_5^1, P_5^2), \sigma_{14} = (P_5^1, P_5^2, P_6^1, P_6^1), \quad \sigma_{15} = (P_6^1, P_6^1), \quad \sigma_{16} = (P_4^1, P_4^2),$$
$$\sigma_{17} = (P_5^1, P_5^2)$$
$$\Psi = (1, 3, 1, 1, 1, 2)$$
$$\Xi = 1$$

$SM = (ST_M, BE_M)$—*the multimodal processes structure*

$$ST_M = (M, mT)$$
$$M = \{mp_1, mp_2\}$$
$$mp_1 = (R_1, R_2, R_3, R_9, R_8, R_{17}, R_{13}),$$
$$mp_2 = (R_{15}, R_{14}, R_{12}, R_8, R_7, R_5, R_6)$$

$mT = \{mt_1, mt_2\}$---*where values of all elements mt_i are equal to* 1 *unit of time*

$$BE_M = (m\Psi, m\Xi)$$
$$m\Psi = (1, 1), m\Xi = 1$$

The next two systems SC^2 and SC^3 differ from SC^1 by different dispatching priority rules employed:

- the priority dispatching rule associated to R_{12} in the SC^2 is as follows:
- $\sigma_{12} = (P_4^1, P_5^1, P_4^2, P_5^2)$,
- the priority dispatching rules associated to R_{12} and R_{14} in the SC^3 are as follows:

$$\sigma_{12} = (P_4^1, P_5^1, P_4^2, P_5^2), \ \sigma_{14} = (P_5^1, P_6^1, P_5^2, P_6^1)$$

The rest of SC^2 and SC^3 parameters are just the same as in the SC^1

The responses to the following questions are sought:

- Do there exist cyclic steady states mSc_1, mSc_2, mSc_3 encompassing behaviors of SC^1, SC^2, SC^3, respectively (i.e. resulting in the cyclic steady states of local and multimodal processes)?
- Is it possible to switch between cyclic steady states mSc_1, mSc_2, mSc_3, i.e. to jump from the one steady cyclic state to another at the same allocation of local cyclic processes, i.e., $mSc_1 \leftrightarrow mSc_2 \leftrightarrow mSc_3$,?

The response to above questions assumes the not empty cyclic steady states space there exists. Let us remind, that a decidability of the relevant reachability problem follows from its Diophantine nature (Smart Nigiel 1998). In general case this problem is NP-hard. The approach aimed at a space of feasible states generation (see Fig. 4) is time expensive (its generation is a NP-hard problem) and results mainly in a deadlock or leading to a deadlock states. An alternative approach, based on CSP formulation CS (9) enables to focus (and to generate if that is possible) on dedicated cyclic steady states both of local as well as multimodal processes.

4.1 Local Processes Scheduling

Consider the SCCP modeling the SC^1 and its SL level.

Searching for a possible cyclic steady state of local processes formulated in terms of CSP can be stated as following constraints satisfaction problem:

$$CS(SC^1) = ((\{X, Tc\}, \{D_X, D_{Tc}\}), C) \tag{14}$$

where:

$X = \{X_1^1, X_2^1, X_3^1, X_4^1, X_4^2, X_5^1, X_5^2, X_6^1\}$

$X_1^1 = \left(x_{1,1,1}^1, x_{1,2,1}^1, x_{1,3,1}^1, x_{1,4,1}^1\right)$ - - - the moments of beginnings of operations executed along the $p_1^1 = (R_1, R_2, R_3, R_4)$

$X_2^1 = \left(x_{2,1,1}^1, x_{2,2,1}^1, x_{2,3,1}^1, x_{2,1,2}^1, x_{2,2,2}^1, x_{2,3,2}^1, x_{2,1,3}^1, x_{2,2,3}^1, x_{2,3,3}^1\right),$

$X_3^1 = \left(x_{3,1,1}^1, x_{3,2,1}^1, x_{3,3,1}^1, x_{3,4,1}^1, x_{3,5,1}^1\right),$

$X_4^1 = \left(x_{4,1,1}^1, x_{4,2,1}^1, x_{4,3,1}^1, x_{4,4,1}^1, x_{4,5,1}^1, x_{4,6,1}^1\right),$

$X_4^2 = \left(x_{4,1,1}^2, x_{4,2,1}^2, x_{4,3,1}^2, x_{4,4,1}^2, x_{4,5,1}^2, x_{4,6,1}^2\right),$

$X_5^1 = \left(x_{5,1,1}^1, x_{5,2,1}^1, x_{5,3,1}^1, x_{5,4,1}^1, x_{5,5,1}^1\right),$

$X_5^2 = \left(x_{5,1,1}^2, x_{5,2,1}^2, x_{5,3,1}^2, x_{5,4,1}^2, x_{5,5,1}^2\right),$

$X_6^1 = \left(x_{6,1,1}^1, x_{6,2,1}^1, x_{6,3,1}^1, x_{6,1,2}^1, x_{6,2,2}^1, x_{6,3,2}^1\right)$

C—the set of constraints consists (due to (11)) of constraints form Table 4.

The solution to the CS (14) implemented in OzMozart platform (on Intel Core Duo 3.00 GHz, 4.00 GB RAM computer, and obtained in the time less than 1 s) is shown in the Table 5.

Therefore, the period Tc_1 of the cyclic steady state obtained is equal to 19. The moments of operations beginning in local processes are shown in the Table 5. The

Table 4 Constraints determining the moments $x^k_{i,j,q}$ of operations beginning for the SC^1 modelled by the SCCP from Fig. 2

Local process P^1_1	Local process P^1_2
$x^1_{1,1,1} = max\left\{\left(x^1_{1,4,1} + t^1_{1,4} - Tc\right); \left(x^1_{1,4,1} + 1 - Tc\right)\right\}$	$x^1_{2,1,1} = max\left\{\left(x^1_{2,3,3} + t^1_{2,3} - Tc\right); \left(x^1_{1,1,1} + 1 + Tc\right)\right\}$
$x^1_{1,2,1} = max\left\{\left(x^1_{1,1,1} + t^1_{1,1}\right); \left(x^2_{4,2,1} + 1\right)\right\}$	$x^1_{2,2,1} = max\left\{\left(x^1_{2,1,1} + t^1_{2,1}\right); \left(x^1_{3,1,1} + 1\right)\right\}$
$x^1_{1,3,1} = max\left\{\left(x^1_{1,2,1} + t^1_{1,2}\right); \left(x^1_{3,1,1} + 1\right)\right\}$	$x^1_{2,3,1} = max\left\{\left(x^1_{2,2,1} + t^1_{2,2}\right); \left(x^1_{2,1,1} + 1\right)\right\}$
$x^1_{1,4,1} = max\left\{\left(x^1_{1,3,1} + t^1_{1,3}\right); \left(x^1_{2,2,3} + 1\right)\right\}$	$x^1_{2,1,2} = max\left\{\left(x^1_{2,3,1} + t^1_{2,3}\right); \left(x^1_{2,2,1} + 1\right)\right\}$
	$x^1_{2,2,2} = max\left\{\left(x^1_{2,1,2} + t^1_{2,1}\right); \left(x^1_{2,3,1} + 1\right)\right\}$
	$x^1_{2,3,2} = max\left\{\left(x^1_{2,2,2} + t^1_{2,2}\right); \left(x^1_{2,1,2} + 1\right)\right\}$
	$x^1_{2,1,3} = max\left\{\left(x^1_{2,3,2} + t^1_{2,3}\right); \left(x^1_{2,2,2} + 1\right)\right\}$
	$x^1_{2,2,3} = max\left\{\left(x^1_{2,1,3} + t^1_{2,1}\right); \left(x^1_{2,3,2} + 1\right)\right\}$
	$x^1_{2,3,3} = max\left\{\left(x^1_{2,2,3} + t^1_{2,2}\right); \left(x^1_{2,1,3} + 1\right)\right\}$
Local process P^1_3	Local process P^1_4
$x^1_{3,1,1} = max\left\{\left(x^1_{3,5,1} + t^1_{3,5} - Tc\right); \left(x^1_{1,4,1} + 1\right)\right\}$	$x^1_{4,1,1} = max\left\{\left(x^1_{4,6,1} + t^1_{4,6} - Tc\right); \left(x^1_{1,3,1} + 1\right)\right\}$
$x^1_{3,2,1} = max\left\{\left(x^1_{3,1,1} + t^1_{3,1}\right); \left(x^2_{4,1,1} + 1\right)\right\}$	$x^1_{4,2,1} = max\left\{\left(x^1_{4,1,1} + t^1_{4,1}\right); \left(x^2_{4,3,1} + 1 - Tc\right)\right\}$
$x^1_{3,3,1} = max\left\{\left(x^1_{3,2,1} + t^1_{3,2}\right); \left(x^2_{5,2,1} + 1\right)\right\}$	$x^1_{4,3,1} = max\left\{\left(x^1_{4,2,1} + t^1_{4,2}\right); \left(x^2_{4,4,1} + 1 - Tc\right)\right\}$
$x^1_{3,4,1} = max\left\{\left(x^1_{3,3,1} + t^1_{3,3}\right); \left(x^1_{3,5,1} + 1 - Tc\right)\right\}$	$x^1_{4,4,1} = max\left\{\left(x^1_{4,3,1} + t^1_{4,3}\right); \left(x^1_{6,2,2} + 1 - Tc\right)\right\}$
$x^1_{3,5,1} = max\left\{\left(x^1_{3,4,1} + t^1_{3,4}\right); \left(x^1_{2,3,3} + 1\right)\right\}$	$x^1_{4,5,1} = max\left\{\left(x^1_{4,4,1} + t^1_{4,4}\right); \left(x^2_{5,2,1} + 1\right)\right\}$
	$x^1_{4,6,1} = max\left\{\left(x^1_{4,5,1} + t^1_{4,5}\right); \left(x^1_{3,3,1} + 1\right)\right\}$

(continued)

Table 4 (continued)

Local process P_4^2	Local process P_5^1
$x_{4,1,1}^2 = max\left\{\left(x_{4,6,1}^2 + t_{4,6}^2 - Tc\right); \left(x_{4,2,1}^1 + 1\right)\right\}$	$x_{5,1,1}^1 = max\left\{\left(x_{5,5,1}^1 + t_{5,5}^1 - Tc\right); \left(x_{3,4,1}^1 + 1\right)\right\}$
$x_{4,2,1}^2 = max\left\{\left(x_{4,1,1}^2 + t_{4,1}^2\right); \left(x_{4,3,1}^1 + 1\right)\right\}$	$x_{5,2,1}^1 = max\left\{\left(x_{5,1,1}^1 + t_{5,1}^1\right); \left(x_{5,3,1}^2 + 1 - Tc\right)\right\}$
$x_{4,3,1}^2 = max\left\{\left(x_{4,2,1}^2 + t_{4,2}^2\right); \left(x_{4,4,1}^1 + 1\right)\right\}$	$x_{5,3,1}^1 = max\left\{\left(x_{5,2,1}^1 + t_{5,2}^1\right); \left(x_{5,4,1}^2 + 1 - Tc\right)\right\}$
$x_{4,4,1}^2 = max\left\{\left(x_{4,3,1}^2 + t_{4,3}^2\right); \left(x_{6,2,1}^1 + 1\right)\right\}$	$x_{5,4,1}^1 = max\left\{\left(x_{5,3,1}^1 + t_{5,3}^1\right); \left(x_{6,1,1}^1 + 1\right)\right\}$
$x_{4,5,1}^2 = max\left\{\left(x_{4,5,1}^2 + t_{4,5}^2\right); \left(x_{4,6,1}^1 + 1\right)\right\}$	$x_{5,5,1}^1 = max\left\{\left(x_{5,4,1}^1 + t_{5,4}^1\right); \left(x_{4,6,1}^2 + 1\right)\right\}$
$x_{4,6,1}^2 = max\left\{\left(x_{4,5,1}^2 + t_{4,5}^2\right); \left(x_{4,1,1}^1 + 1 + Tc\right)\right\}$	
Local process P_5^2	Local process P_6^1
$x_{5,1,1}^2 = max\left\{\left(x_{5,5,1}^2 + t_{5,5}^2 - Tc\right); \left(x_{5,2,1}^1 + 1\right)\right\}$	$x_{6,1,1}^1 = max\left\{\left(x_{6,3,2}^1 + t_{6,3}^1 - Tc\right); \left(x_{4,5,1}^2 + 1\right)\right\}$
$x_{5,2,1}^2 = max\left\{\left(x_{5,1,1}^2 + t_{5,1}^2\right); \left(x_{5,3,1}^1 + 1\right)\right\}$	$x_{6,2,1}^1 = max\left\{\left(x_{6,1,1}^1 + t_{6,1}^1\right); \left(x_{6,3,2}^1 + 1 - Tc\right)\right\}$
$x_{5,3,1}^2 = max\left\{\left(x_{5,1,1}^2 + t_{5,1}^2\right); \left(x_{5,4,1}^1 + 1\right)\right\}$	$x_{6,3,1}^1 = max\left\{\left(x_{6,2,1}^1 + t_{6,2}^1\right); \left(x_{5,5,1}^2 + 1\right)\right\}$
$x_{5,4,1}^2 = max\left\{\left(x_{5,3,1}^2 + t_{5,3}^2\right); \left(x_{5,5,1}^1 + 1\right)\right\}$	$x_{6,1,2}^1 = max\left\{\left(x_{6,3,1}^1 + t_{6,3}^1\right); \left(x_{6,2,1}^1 + 1\right)\right\}$
$x_{5,5,1}^2 = max\left\{\left(x_{5,4,1}^2 + t_{5,4}^2\right); \left(x_{5,1,1}^1 + 1 + Tc\right)\right\}$	$x_{6,2,2}^1 = max\left\{\left(x_{6,1,2}^1 + t_{6,1}^1\right); \left(x_{6,3,1}^1 + 1\right)\right\}$
	$x_{6,3,2}^1 = max\left\{\left(x_{6,2,2}^1 + t_{6,2}^1\right); \left(x_{6,4,1}^1 + 1\right)\right\}$

Table 5 The moments $x^k_{i,j,q}$ of operations beginning for SCCP from Fig. 2

Start moments	$x^1_{1,1,1}$, $x^1_{3,5,1}$, $x^1_{5,3,1}$, $x^1_{6,2,2}$	$x^1_{2,1,1}$, $x^2_{5,2,1}$, $x^1_{6,3,2}$, $x^1_{4,4,1}$	$x^1_{1,2,1}$, $x^1_{4,5,1}$, $x^2_{4,3,1}$	$x^1_{3,1,1}$, $x^1_{6,1,1}$	$x^1_{2,2,1}$, $x^1_{5,4,1}$, $x^1_{6,2,1}$	$x^2_{4,4,1}$, $x^2_{5,3,1}$	$x^1_{2,3,1}$, $x^1_{3,2,1}$	$x^1_{1,3,1}$	$x^1_{2,1,2}$	$x^1_{3,3,1}$
Values	0	1	2	3	4	5	6	7	8	9
Corresponding states	Sl^0	Sl^1	Sl^2	Sl^3	Sl^4	Sl^5	Sl^6	Sl^7	Sl^8	Sl^9
Start moments	$x^1_{2,2,2}$, $x^1_{4,6,1}$	$x^1_{4,1,1}$, $x^2_{4,5,1}$	$x^1_{2,3,2}$, $x^1_{3,4,1}$, $x^1_{4,2,1}$, $x^2_{4,6,1}$	$x^1_{4,3,1}$, $x^2_{4,1,1}$, $x^1_{5,5,1}$	$x^1_{2,1,3}$, $x^2_{4,2,1}$, $x^2_{5,4,1}$	$x^1_{5,1,1}$	$x^1_{2,2,3}$, $x^2_{5,5,1}$	$x^1_{1,4,1}$, $x^1_{5,2,1}$, $x^1_{6,3,1}$	$x^1_{2,3,3}$, $x^2_{5,1,1}$, $x^1_{6,1,2}$	
Values	10	11	12	13	14	15	16	17	18	
Corresponding states	Sl^{10}	Sl^{11}	Sl^{12}	Sl^{13}	Sl^{14}	Sl^{15}	Sl^{16}	Sl^{17}	Sl^{18}	

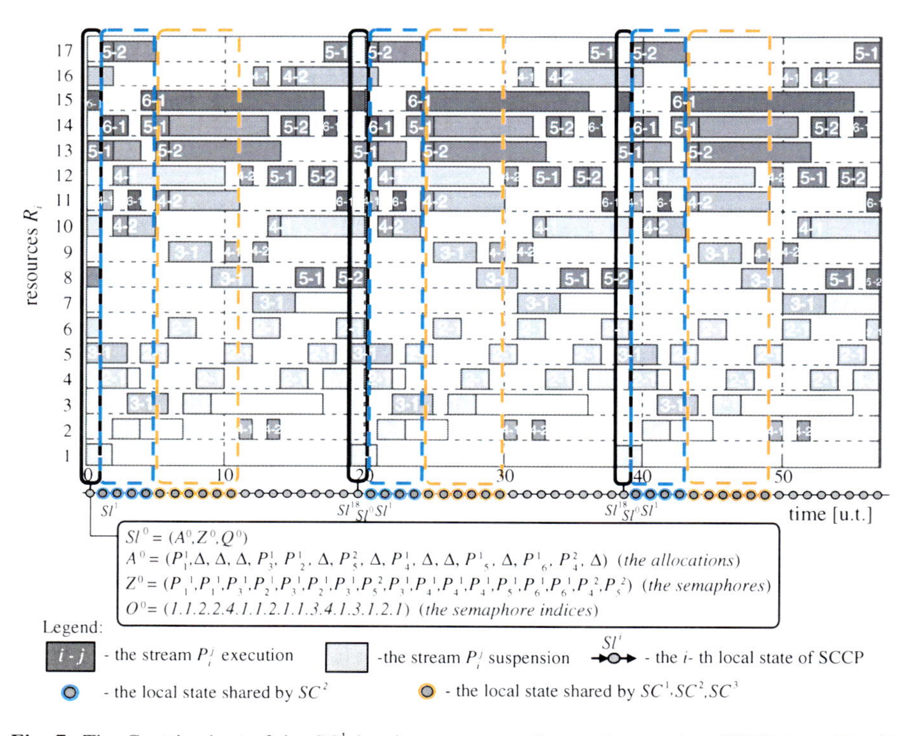

Fig. 7 The Gantt's chart of the SC^1 local processes cyclic steady state (see SCCP from Fig. 2)

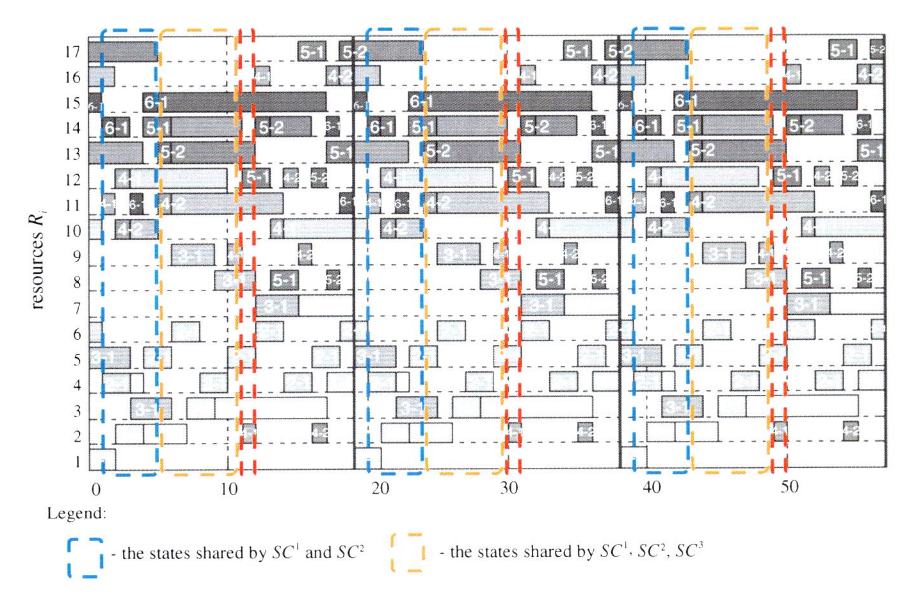

Fig. 8 The Gantt's chart of the SC^2 local processes cyclic steady state (see SCCP from Fig. 2)

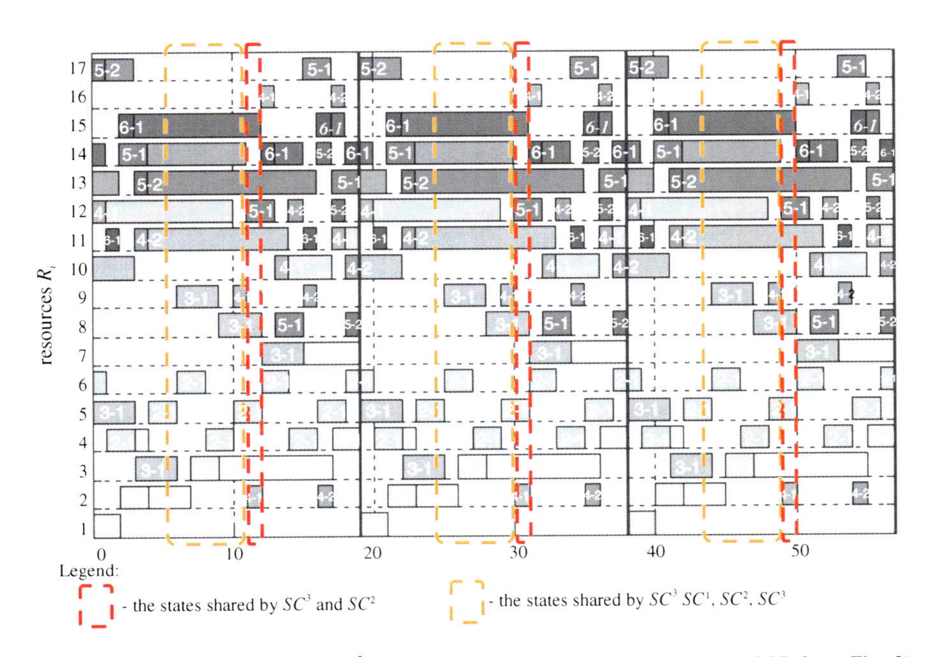

Fig. 9 The Gantt's chart of the SC^3 local processes cyclic steady state (see SCCP from Fig. 2)

Table 6 Constraints determining the moments $mx_{i,j,k}$ of the SC^1 operations beginning

Multimodal process mP_1

$$mx_{1,1,1} = min\left\{\left(x^1_{1,1,1} + a \cdot Tc\right) | a \in \mathbb{Z}, x^1_{1,1,1} + a \cdot Tc \geq mx_{1,7,1} + mt_{1,7,1} - Tm\right\},$$

$$mx_{1,2,1} = min\left\{\left(x^1_{1,2,1} + a \cdot Tc\right) | a \in \mathbb{Z}, x^1_{1,2,1} + a \cdot Tc \geq mx_{1,1,1} + mt_{1,1,1}\right\},$$

$$mx_{1,3,1} = min\left\{\left(x^1_{1,3,1} + a \cdot Tc\right) | a \in \mathbb{Z}, x^1_{1,3,1} + a \cdot Tc \geq mx_{1,2,1} + mt_{1,2,1}\right\},$$

$$mx_{1,4,1} = min\left\{\left(x^1_{3,2,1} + a \cdot Tc\right) | a \in \mathbb{Z}, x^1_{3,2,1} + a \cdot Tc \geq mx_{1,3,1} + mt_{1,3,1}\right\},$$

$$mx_{1,5,1} = min\left\{\left(x^1_{3,3,1} + a \cdot Tc\right) | a \in \mathbb{Z}, x^1_{3,3,1} + a \cdot Tc \geq mx_{1,4,1} + mt_{1,4,1}\right\},$$

$$mx_{1,6,1} = min\left\{\left(x^1_{5,2,1} + a \cdot Tc\right), \left(x^2_{5,2,1} + a \cdot Tc\right) | a \in \mathbb{Z}, x^1_{5,2,1} + a \cdot Tc \geq mx_{1,5,1} \right.$$
$$\left. mt_{1,5,1}, x^2_{5,2,1} + a \cdot Tc \geq mx_{1,5,1} + mt_{1,5,1}\right\}$$

$$mx_{1,7,1} = min\left\{\left(x^1_{5,3,1} + a \cdot Tc\right), \left(x^2_{5,3,1} + a \cdot Tc\right) | a \in \mathbb{Z}, x^1_{5,2,1} + a \cdot Tc \right.$$
$$\left. \geq mx_{1,5,1} + mt_{1,5,1}, x^2_{5,2,1} + a \cdot Tc \geq mx_{1,6,1} + mt_{1,6,1}\right\}$$

Multimodal process mP_2

$$mx_{2,1,1} = min\left\{\left(x^1_{6,3,1} + a \cdot Tc\right), \left(x^1_{6,3,2} + a \cdot Tc\right) | a \in \mathbb{Z}, x^1_{6,3,1} + a \cdot Tc \geq mx_{2,7,1} \right.$$
$$\left. + mt_{2,7,1} - Tm, x^1_{6,3,2} + a \cdot Tc \geq mx_{2,7,1} + mt_{2,7,1} - Tm\right\},$$

$$mx_{2,2,1} = min\left\{\left(x^1_{6,1,1} + a \cdot Tc\right), \left(x^1_{6,1,2} + a \cdot Tc\right) | a \in \mathbb{Z}, x^1_{6,1,1} + a \cdot Tc \geq \right.$$
$$\left. mx_{2,1,1} + mt_{2,1,1}, x^1_{6,1,2} + a \cdot Tc \geq mx_{2,1,1} + mt_{2,1,1}\right\}$$

$$mx_{2,3,1} = min\left\{\left(x^1_{5,5,1} + a \cdot Tc\right), \left(x^2_{5,5,1} + a \cdot Tc\right) | a \in \mathbb{Z}, x^1_{5,5,1} + a \cdot Tc \geq \right.$$
$$\left. mx_{2,2,1} + mt_{2,2,1}, x^2_{5,5,1} + a \cdot Tc \geq mx_{2,2,1} + mt_{2,2,1}\right\}$$

$$mx_{2,4,1} = min\left\{\left(x^1_{5,1,1} + a \cdot Tc\right), \left(x^2_{5,1,1} + a \cdot Tc\right) | a \in \mathbb{Z}, x^1_{5,1,1} + a \cdot Tc \geq \right.$$
$$\left. mx_{2,3,1} + mt_{2,3,1}, x^2_{5,1,1} + a \cdot Tc \geq mx_{2,3,1} + mt_{2,3,1}\right\}$$

$$mx_{2,5,1} = min\left\{\left(x^1_{3,4,1} + a \cdot Tc\right) | a \in \mathbb{Z}, x^1_{3,4,1} + a \cdot Tc \geq mx_{2,4,1} + mt_{2,4,1}\right\},$$

$$mx_{2,6,1} = min\left\{\left(x^1_{3,5,1} + a \cdot Tc\right) | a \in \mathbb{Z}, x^1_{3,5,1} + a \cdot Tc \geq mx_{2,5,1} + mt_{2,5,1}\right\},$$

$$mx_{2,7,1} = min\left\{\left(x^1_{2,3,1} + a \cdot Tc\right), \left(x^1_{2,3,2} + a \cdot Tc\right), \left(x^1_{2,3,3} + a \cdot Tc\right) | a \in \mathbb{Z}, x^1_{2,3,1} + a \cdot Tc. \geq \right.$$
$$\left. mx_{2,6,1} + mt_{2,6,1}, x^1_{2,3,2} + a \cdot Tc \geq mx_{2,6,1} + mt_{2,6,1}, x^1_{2,3,3} + a \cdot Tc \geq mx_{2,6,1} + mt_{2,6,1}\right\}$$

Gantt's chart of the cyclic steady state behavior is shown in Fig. 7. Let us note, that each unit time corresponds to a system state, i.e. processes allocation to resources. So, the cyclic steady state of the SC^1 consists of 19 states: $Sc_1 = \{Sl^0, Sl^1, \ldots, Sl^{18}\}$, and (similar to the Fig. 4) the states Sl^i encompassing succeeding allocations are distinguished by "O" (see Fig. 7).

In the same way the Gantt's charts for SC^2 and SC^3 can be determined, see Figs. 8 and 9. Both systems have the same period equal to 19 (i.e., the same as for the SC^1).

4.2 Multimodal Processes Scheduling

The cyclic steady state obtained provide the response to the first from earlier mentioned questions, i.e. the systems SC^1, SC^2 and SC^3 encompass the cyclic steady states. The cyclic steady states of local cyclic processes determine the cyclic steady states of the multimodal processes.

Consider the cyclic steady state Sc_1 of SCCP local processes. The question is: What is a cyclic steady state of the multimodal processes mSc_1 executed in this system? Let us assume two multimodal processes mP_1 and mP_2. So, the relevant CSP, i.e. $CS(SC)$ defined as (9), which can be seen in terms of the problem (14) extended by variables $mX = \{mX_1, mX_2\}$, where:

$$\begin{aligned} mX_1 &= \left(mx_{1,1,1}, mx_{1,2,1}, mx_{1,3,1}, mx_{1,4,1}, mx_{1,5,1}, mx_{1,6,1}, mx_{1,7,1}\right) \\ mX_2 &= \left(mx_{2,1,1}, mx_{2,2,1}, mx_{2,3,1}, mx_{2,4,1}, mx_{2,5,1}, mx_{2,6,1}, mx_{2,7,1}\right) \end{aligned}$$

as well as constraints (stated in (13)) specified in the Table 6, has to be solved. The solution obtained consist of the already obtained solution to the problem (14) (see the Table 5) as well as moments mX_1, mX_2 of operations beginning, see the Table 7.

The Gantt's diagram illustration of the cyclic steady state of multimodal processes (composed of 57 states $mSc_1 = \{S^0, S^1, \ldots, S^{57}\}$) behavior is show in Fig. 10. The obtained period is equal to $Tm_1 = 57$ (i.e. a multiplicity of the period $Tc_1 = 19$).

That means that within a period the multimodal processes complete their one execution the local processes complete three times, i.e. local processes period is three times shorter than the multimodal one.

Consider the ith multimodal process completion time denoted Te_i, i.e. the time period between the beginning of the first operation from the ith route, and the end of the last operations in this route. Let us note that the cycle of the cyclic steady state equals to $Tm_1 = 57$ u.t. while the Te_1 (Te_2) for the mP_1 (mP_2) equals to 40 u.t. (46 u.t.), respectively.

In case of systems SC^2 and SC^3 they have the same twice longer cycle time $Tm_2 = Tm_3 = 114$, however mP_1 executes three times, and mP_2 two times, and their completion times are shorter $Te_1 = 38$ u.t. and $Te_2 = 46$ u.t., respectively

Table 7 The moments $mx_{i,j,k}$ of SC^1 operations beginning

Start moments	$mx_{1,1,1}$, $mx_{2,2,1}$	$mx_{1,2,1}$	$mx_{1,3,1}$	$mx_{2,3,1}$	$mx_{2,4,1}$	$mx_{1,4,1}$
Values	0	2	7	15	17	25
Local states	Sl^0	Sl^2	Sl^7	Sl^{15}	Sl^{17}	Sl^6
Multimodal states	S^0	S^2	S^7	S^{15}	S^{17}	S^{25}
Start moments	$mx_{1,5,1}$	$mx_{2,5,1}$	$mx_{1,6,1}$, $mx_{2,6,1}$	$mx_{1,7,1}$	$mx_{2,7,1}$	$mx_{2,1,1}$
Values	28	31	38	40	44	51
Local states	Sl^9	Sl^{12}	Sl^0	Sl^2	Sl^6	Sl^{17}
Multimodal states	S^{28}	S^{31}	S^{38}	S^{40}	S^{40}	S^{51}

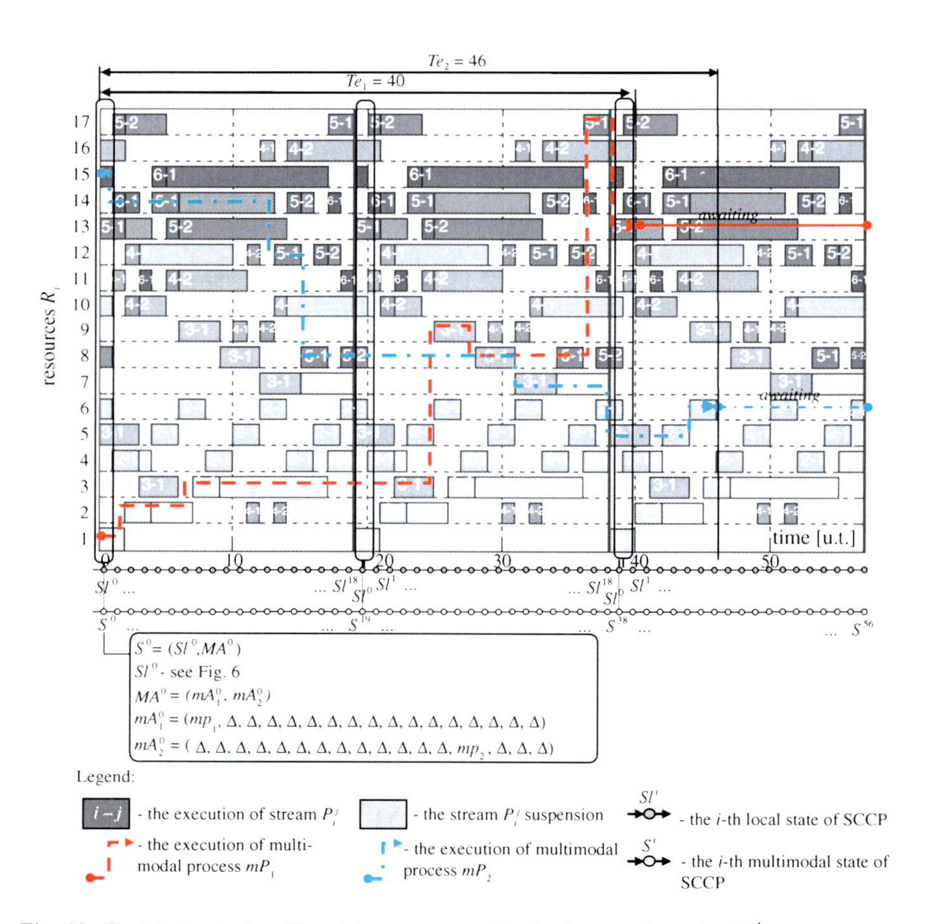

Fig. 10 Gantt's chart of multimodal processes cyclic steady state from the SC^1

(see Figs. 11 and 12). Moreover, $Te_1 = 38$ u.t. and $Te_2 = 44$ u.t for the SC^3, see the Table 8.

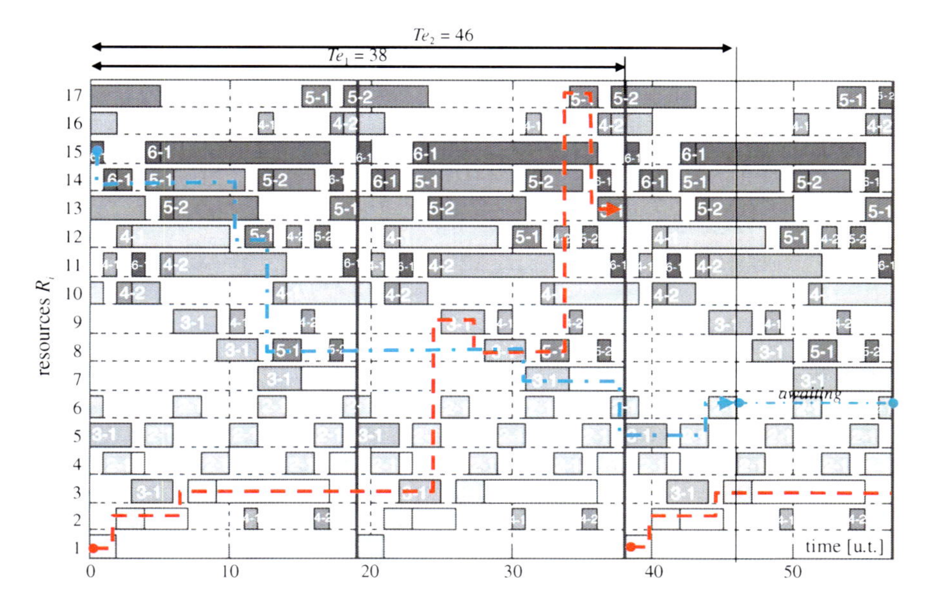

Fig. 11 Gantt's chart of multimodal processes cyclic steady state of the SC^2

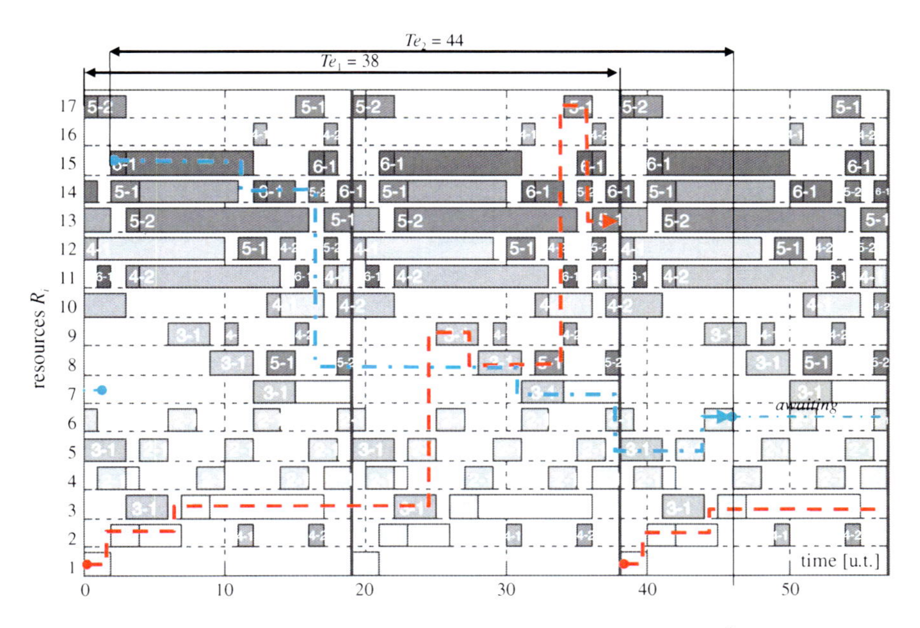

Fig. 12 Gantt's chart of multimodal processes cyclic steady state of the SC^3

Table 8 The cycles and processes completion times for SC^1, SC^2 and SC^3

SC^1	$Tm_1 = 57$ u.t.	$Te_1 = 40$ u.t.	$Te_2 = 46$ u.t.
SC^2	$Tm_2 = 114$ u.t.	$Te_1 = 38$ u.t.	$Te_2 = 46$ u.t.
SC^3	$Tm_3 = 114$ u.t.	$Te_1 = 38$ u.t.	$Te_2 = 44$ u.t.

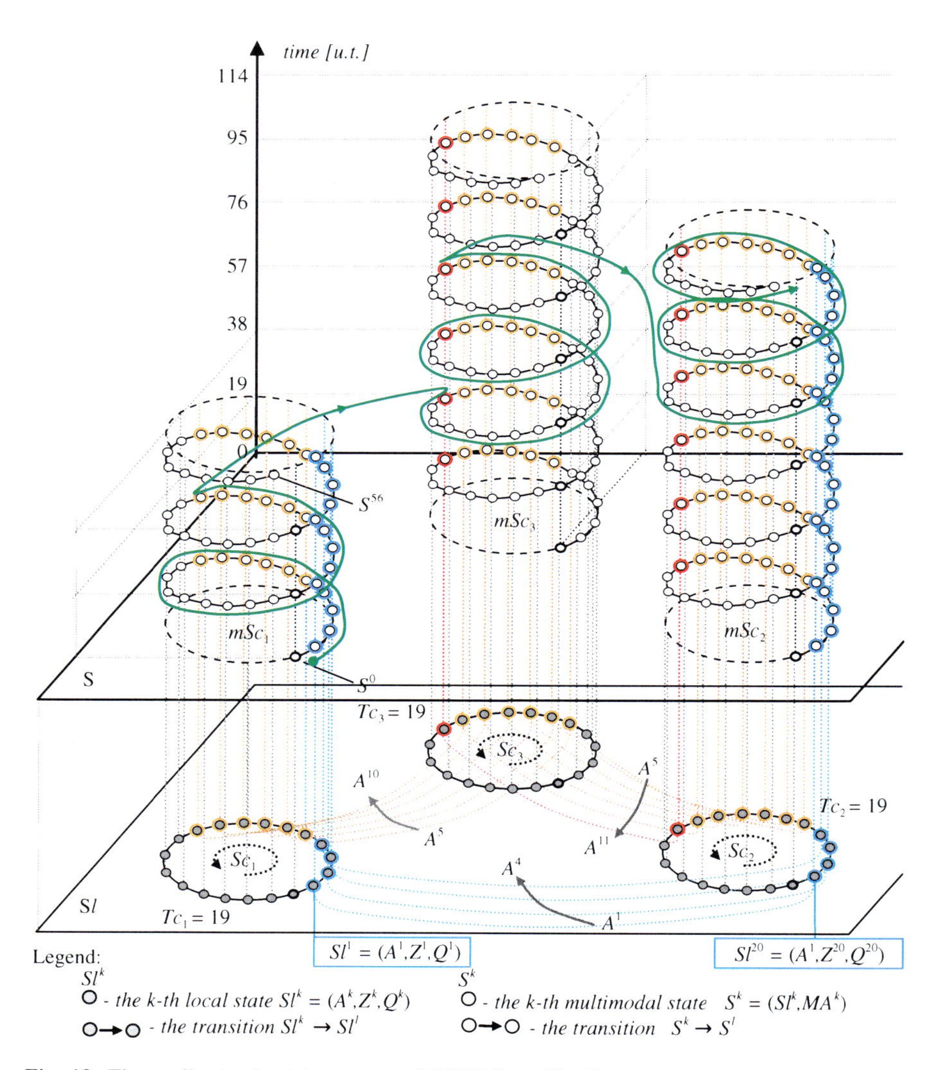

Fig. 13 The cyclic steady states spaces of SCCP from Fig. 2

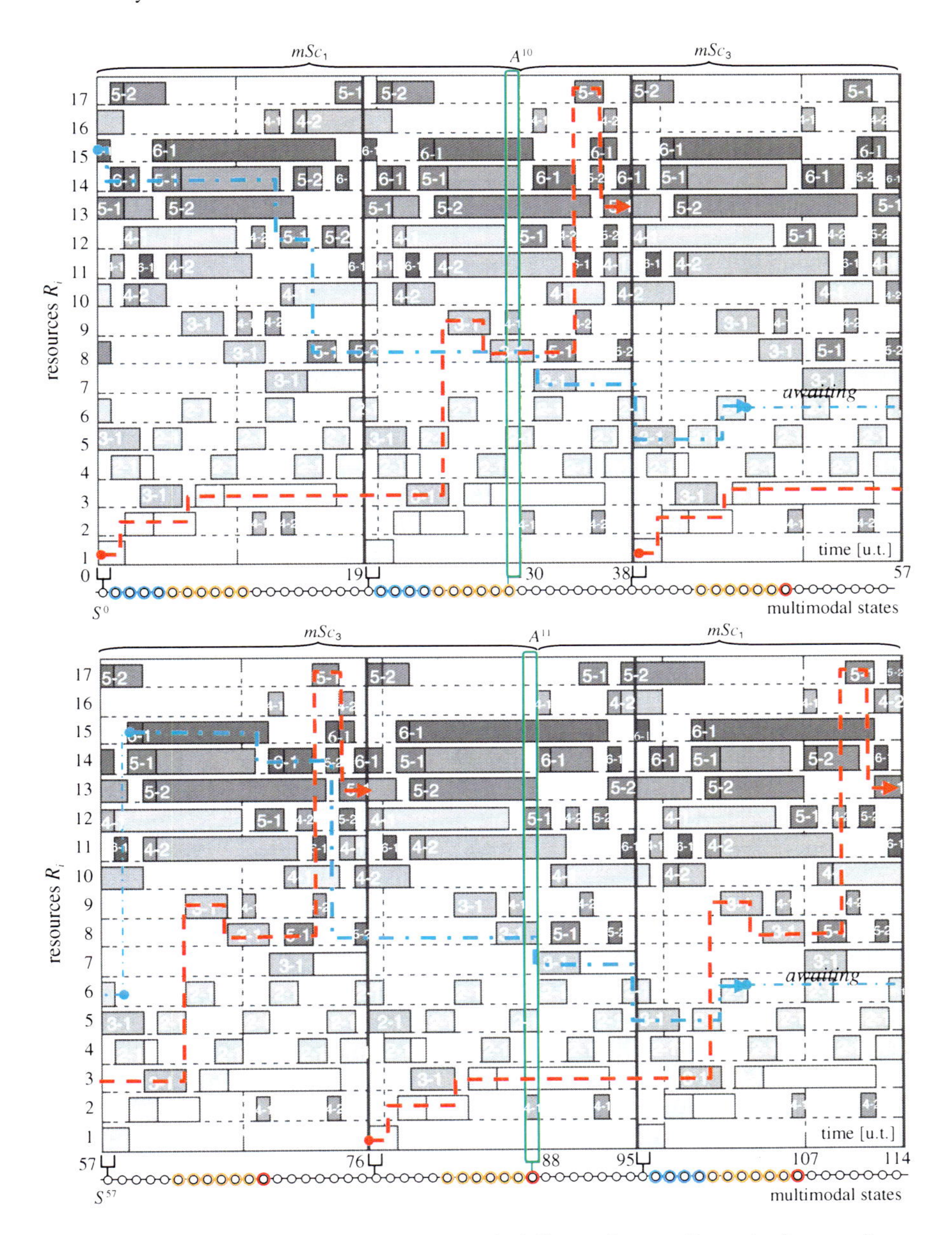

Fig. 14 Gantt's chart of cyclic steady states rescheduling: $mSc_1 \to mSc_3$ and $mSc_3 \to mSc_2$

4.3 Reachability of Cyclic Steady States Space

The cyclic steady states from the Table 8 specify standing behind of them variants in terms of cycles of the local, and the multimodal cyclic processes. In case of cyclic steady states space considered the all states specify the same local processes cycle while differ in multimodal processes cycle values. Consequently, those differences imply different completion times of multimodal processes. That means, the states space allows one to search either for the shortest cycle or for the shortest makespan multimodal schedules.

Besides of multimodal processes scheduling the issue of their rescheduling plays a pivotal role. Due to the property 2 and 3 mentioned see the Sect. 3.1 the direct rescheduling from one cyclic steady state to another is allowed at states possessing the same local processes allocation. For instance, the states sharing the same allocations, (distinguished by different colors: blue, read, and orange) are shown in Figs. 7, 8 and 9. Taking into account such distinguished allocations the sequence of possible reschedules (depicted by bold green line), omitting however the transient periods, is shown in Fig. 13. The local processes cyclic states space is shown on the $\mathbb{S}l$ level and the multimodal processes cyclic states space is show on the $\mathbb{S}$ level.

The states with allocations $A^1 - A^{11}$ shared by different cyclic steady states are distinguished by different colors, see Fig. 13. Allocations $A^1 - A^4$ (distinguished by blue doted lines) are shared by mSc_1 and mSc_2, allocations $A^5 - A^{10}$ (distinguished by orange doted lines) are shared by mSc_1, mSc_2, and mSc_3, and the allocations $A^5 - A^{11}$ (distinguished by red doted lines) are shared by mSc_2 and mSc_3.

That means the rescheduling from a one multimodal processes cyclic steady state to other one is possible only in the states $A^1 - A^{11}$ shared by other cyclic steady states. In other words, to each state from the local processes steady cyclic state correspond the states from multimodal processes steady cyclic state. In case of states sharing allocations by different local processes cyclic steady states the switching between relevant (sharing the same color) multimodal processes cyclic steady states can be obtained by adjusting those states' semaphores Z^i and/or indices Q^i and/or dispatching rules Θ. Illustration of such rescheduling, i.e., switching between mSc_1 and mSc_3 while using the state with allocation A^{10}, and switching between mSc_3 and mSc_2 while using the state A^{11}, is distinguished by the green line, see Fig. 13 with the Gantt's diagram in Fig. 14.

5 Concluding Remarks

The constraint satisfaction problem of cyclic scheduling of local and multimodal concurrently flowing processes is considered. It is assumed that multimodal processes are composed of sequences of local cyclic processes. For instance, in case local processes encompass a network of dedicated AGVs lines the processes responsible for workpieces shifting between destinations points while crossing different lines are multimodal ones.

Since AGVS structure constraints its SCCP behavior, there are two fundamental problems one have to face with: Does there exist a set of dispatching rules subject to AGVS's structure constraints guaranteeing solution to a CSP representation of the cyclic scheduling problem? What set of dispatching rules subject to assumed cyclic behavior of AGVS guarantees solution to a CSP representation of the cyclic scheduling problem?

In that context the chapter's contribution is to propose a declarative approach providing unified framework for local and multimodal processes evaluation while allowing one to take into account both direct and reverse problems formulation of the cyclic scheduling. Moreover, the framework employed enables to evaluate the SCCP behavior on the base of the given processes layout, operation times and dispatching rules employed, especially to compose elementary systems in such a way as to obtain the required quantitative and qualitative behavioral features. Concluding, a method allowing one to replace an exhaustive search for the admissible control by a step-by-step structural design guaranteeing the required system behavior is our main contribution.

References

Bocewicz G, Banaszak Z (2013) Declarative approach to cyclic steady states space refinement: periodic processes scheduling. Int J Adv Manuf Technol Vol. 67, Issue 1–4, Springer, pp 137–155

Bocewicz G, Bach I, Banaszak Z (2009a) Logic-algebraic method based and constraints programming driven approach to AGVs scheduling. In. Int J Intell Inf Database Syst 3(1):56–74

Bocewicz G., Wójcik R., Banaszak Z (2009b) On undecidability of cyclic scheduling problems. In: Mapping relational databases to the semantic web with original meaning. Lecture Notes in Computer Science, LNCS, vol 5914. Springer, Berlin, pp 310–321

Bocewicz G, Wójcik R, Banaszak Z (2011a) Toward cycling scheduling of concurrent multimodal processes. In: Computational collective intelligence: technologies and applications. Lecture Notes in Artificial Intelligence, LNAI, vol 6922. Springer, Berlin, pp 448–457

Bocewicz G, Wójcik R, Banaszak Z (2011b) Cyclic steady state refinement. In: Abraham A, Corchado JM, Rodríguez González S, de Paz Santana JF (eds) International symposium on distributed computing and artificial intelligence. Series: Advances in intelligent and soft computing, vol 91. Springer, Berlin, pp 191–198

Cai X, Li KN (2000) A genetic algorithm for scheduling staff of mixed skills under multi-criteria. Eur J Oper Res 125:359–369

Dang QV, Nielsen I, Steger-Jensen K (2011) Scheduling a single mobile robot for feeding tasks in a manufacturing cell. International conference advances in production management systems, Stavanger, Norway, 2011

Gaujal B, Jafari M, Baykal-Gursoy M, Alpan G (1995) Allocation sequences of two processes sharing a resource. IEEE Trans Robotics Autom 11(5):748–753

Guy RK (1994) Diophantine equations. Ch. D in unsolved problems in number theory, 2nd edn. Springer, New York, pp 139–198

Lawley MA, Reveliotis SA, Ferreira PM (1998) A correct and scalable deadlock avoidance policy for flexible manufacturing systems. IEEE Trans Robotics Autom 14(5):796–809
Levner E, Kats V, Alcaide D, Pablo L, Cheng TCE (2010) Complexity of cyclic scheduling problems: a state-of-the-art survey. Comput Ind Eng 59(2):352–361
Liebchen C, Möhring RH (2002) A case study in periodic timetabling. Electron Notes Theor Comput Sci 66(6):21–34
Pinedo ML (2005) Planning and scheduling in manufacturing and services. Springer, New York
Polak M, Majdzik P, Banaszak Z, Wójcik R (2004) The performance evaluation tool for automated prototyping of concurrent cyclic processes. Fundamenta Informaticae 60(1-4):269–289
Schulte CH, Smolka G, Wurtz J (1998) Finite domain constraint programming in Oz, DFKI OZ documentation series, German Research Center for Artificial Intelligence, Saarbrucken, Germany, 1998
Smart Nigiel P (1998) The algorithmic resolution of Diophantine equations. London Mathematical Society Student Text, 41. Cambridge University Press, Cambridge
Song J-S, Lee T-E (1998) Petri net modeling and scheduling for cyclic job shops with blocking. Comput Ind Eng 34(2):281–295
Steger-Jensen K, Hvolby HH, Nielsen P, Nielsen I (2011) Advanced planning & scheduling technology. Prod. Plan Control 22(8):800–808
Trouillet B, Korbaa O, Gentina J-CK (2007) Formal approach for FMS cyclic scheduling. IEEE SMC Trans Part C 37(1):126–137
Von Kampmeyer T (2006) Cyclic scheduling problems. Ph.D. dissertation, Fachbereich Mathematik/Informatik, Universität Osnabrück
Wang B, Yang H, Zhang Z-H (2007) Research on the train operation plan of the Beijing-Tianjin inter-city railway based on periodic train diagrams. Tiedao Xuebao/J China Railway Soc 29(2):8–13
Wójcik R (2007) Constraint programming approach to designing conflict-free schedules for repetitive manufacturing processes. In: Cunha PF, Maropoulos PG (eds) Digital enterprise technology. Perspectives and future challenges. Springer, New York, pp 267–274

Decision Support in Automotive Supply Chain Management: Declarative and Operational Research Approach

Paweł Sitek and Jarosław Wikarek

Abstract This chapter presents the two-phase decision support concepts in automotive supply chain. The proposed concept applied two environments-mathematical programming (operation research approach) and constraint logic programming (declarative approach). We present an approach that allows modeling and solution of the same decision making model in both environments independently but also to communicate between them. In this chapter the decision making model as a problem of optimizing the cost in supply chain under resources, multimodal and environmental constraints in the form of MILP (Mixed Integer Linear Programming) has been presented. The numerical experiments were carried out using sample data to show the possibilities of practical decision support and optimization of the automotive supply chain.

Keywords Decision support · Supply chain · Environmental constraints · Optimization · MILP · CLP

1 Introduction

Peter Drucker described the automotive industry as "the industry of industries," because it consumes output from just about every other manufacturing industry (Drucker 2003). The reason for this is the modern vehicle. The typical vehicle is made up of approximately more than ten thousand detailed parts with about one

P. Sitek (✉) · J. Wikarek
Kielce University of Technology, Institute of Management Control Systems, Al. 1000-lecia PP 7, 25-314 Kielce, Poland
e-mail: sitek@tu.kielce.pl

J. Wikarek
e-mail: j.wikarek@tu.kielce.pl

P. Golinska (ed.), *Environmental Issues in Automotive Industry*, EcoProduction, DOI: 10.1007/978-3-642-23837-6_7,

thousand key components coming together at assembly. The automotive industry consumes a significant percentage of the world's output of rubber, malleable iron, machine tools, glass, semiconductors, aluminum, steel, plastic and textiles etc. The market for new motor vehicles was approximately 78 million units in 2010. This includes passenger cars (78 %), light commercial vehicles (including sport utility vehicles (SUVs) and pickups; 18 %), and commercial vehicles and buses (4 %) (Kelly 2012). The scale of the problems, risks and impacts associated with the automotive industry is huge. Alongside economic issues associated with shortening production cycles, shortening supply cycles, increasing productivity, etc., there is a problem of impact of the entire automotive industry on the environment.

The key issue appears to be reaching optimal or feasible decisions continuously under the conditions of changing time, cost, performance and environmental constraints.

Since the problems and constraints mentioned above concern nearly all participants and sections of the supply chain to varying degrees or the whole chain, it is vital to develop such models and methodologies that will easily allow providing decision support for the supply chain management.

The chapter presents the concept of decision support system for the supply chain management, subject to certain groups of constraints (relating to cost, time, modes of transport and environment) that occur throughout the process of decision-making. The concept is illustrated by examples of decision-making models together with their implementation and numerical examples.

2 Supply Chain in Automotive Industry

The issue of the supply chain is the area of science and practice that has been strongly developing since the 1980s of the last century. Numerous definitions and a supply chain reference model have been developed (Simchi-Levi et al. 2003; Shapiro 2001). The supply chain is commonly seen as a collection of various types of companies (raw materials, production, trade, logistics, etc.) working together to improve the flow of products, information and finance. As the words in the term indicate, the supply chain is a combination of its individual links in the process of supplying products (material and services) to the market.

Huang et al. (Huang et al. 2003) studied the shared information of supply chain production. They considered and proposed four classification criteria: supply chain structure, decision level, modeling approach and shared information.

Supply chain structure: It defines the way various organizations within the supply chain are arranged and related to each other. The supply chain structure falls into four main types (Beamon and Chen 2001): Convergent: each node in the chain has at least one successor and several predecessors. Divergent: each node has one predecessor, at least, and several successors. Conjoined, which is a combination of each convergent chain and one divergent chain. Network: this cannot be classified as convergent, divergent or conjoined, and is more complex than the three previous types.

Decision level: Three decision levels may be distinguished in terms of the decision to be made: strategic, tactical and operational and its corresponding period, i.e., long-term, mid-term and short-term.

Supply chain analytical modeling approach: This approach consists in the type of representation, in this case, mathematical relationships, and the aspects to be considered in the supply chain. Most literature describes and discusses the linear programming-based modeling approach, mixed integer linear programming models in particular (Kanyalkar and Adil 2005; Perea-lopez et al. 2003; Park 2005; Jung et al. 2008; Rizk et al. 2006).

Shared information: This consists in the information shared between each network node determined by the model, which enables production, distribution and transport planning in accordance with the purpose drawn up. The shared information process is vital for effective supply chain production, distribution and transport planning. In terms of centralized planning, this information flows from each node of the network where the decisions are made. Shared information includes the following groups of parameters: resources, inventory, production, transport, demand, etc. Minimization of total costs is the main purpose of the models presented in the literature (Rizk et al. 2006; Selim et al. 2008; Lee and Kim 2000; Chern and Hsieh 2007; Jang et al. 2002) while maximization of revenues or sales is considered to a smaller scale (Park 2005; Timpe and Kallrath 2000).

The industry supply chain in the automotive industry stretches from the producers of raw materials through to the assembly of the most sophisticated electronic, automatic and computing technologies (Tang and Qian 2008). The major components of the supply chain include suppliers and warehouses (tier 1–3 or 1–2), OEMs, distribution centre's, dealers, customers. Most automotive OEMs create 30–35 % of value internally and delegate the rest to their suppliers (Dietz et al. 2004). Example of a typical two-tier supply chain with multimodal transport for the automotive industry is shown in Fig. 1.

The automotive industry has undergone a transformational evolution over the last two decades. The traditional method for designing an automotive supply chain requires a fully integrated, lean materials flow pipeline, certain design constructs and activities have to be engineered into the supply chain (Hugo et al. 2004).

Historically, the automotive industry operated under a "push" model, where input data worked out huge and expensive marketing and sales departments in the form of plans to sell specific models, etc.

With the rapid growth of the Internet, the data became easily available to both producers and consumers of vehicles (Tang and Qian 2008). This resulted in the industry's primary focus on lean, "Just-In-Time" manufacturing processes and their supporting technologies. The participants in the supply chain (OEMs, suppliers, logistic providers etc.) invest a lot of money in re-engineering processes and technologies to support a demand-driven model.

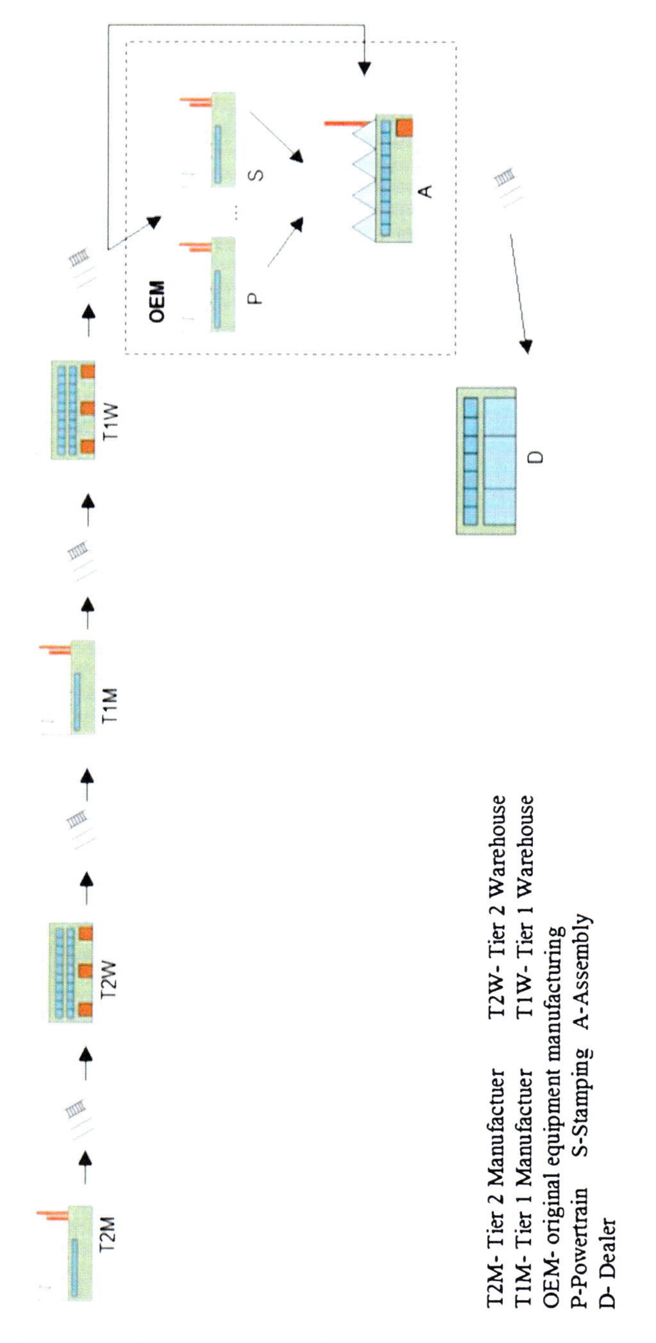

Fig. 1 Exemplified automotive supply chain with multimode transport

Due to the high price of the reengineering and supporting technologies, for instance, ERP and MRP II systems, efforts were limited to selected participants of supply chain.

At the same time, techniques such as lean manufacturing and "Just-In-Time" keep inventory levels low and require frequent re-supply throughout the supply chain—which further adds to the energy consumption and emissions, depending on the product, and has a considerable impact on the environment.

Supply chains in automotive industry are strongly affected by increasing energy prices and greenhouse gas emission constraints.

Global supply chains now span long distances and require significant use of fossil fuels to deliver goods to consumers. In addition, the low inventory levels due to the adoption of lean principles require frequent replenishment of goods throughout a supply chain. This can further increase the energy use and emissions, depending on the product Venkat and Wakeland (2006).

The resource intensive consumption nature of supply-chain networks is highlighted by the fact that transport now consumes nearly a quarter of all the petroleum worldwide and produces over 10 % of the carbon emissions from fossil fuels (Heywood 2006).

There are many metrics to measure the environmental performance of a supply chain as materials move through transport links and are stored temporarily in storage facilities along the way. The most important are Venkat (2007):

- Energy or fuel consumption due to transport and storage.
- Carbon-dioxide (CO_2) emissions due to transport and storage.
- Financial cost of operating the supply chain

3 Motivation

Simultaneously considering the supply chain production, distribution processes in distribution centers and transport-planning problems greatly advances the efficiency of all processes. The literature in the field is vast, so an extensive review of existing research on the topic is extremely helpful in modeling and research. Comprehensive surveys on these problems and their generalizations were published, for example, in (Huang et al. 2003).

In our approach, we are considering a case of the supply chain where:

- the environmental aspects of use of transport modes (depend on the use of fossil fuels and carbon-dioxide emissions etc.);
- the shared information process in the supply chain consists of resources (capacity, versatility, costs), inventory (capacity, versatility, costs, time), production (capacity, versatility, costs), product (volume), transport (cost, mode, time), demand, etc;
- the transport is multimodal (several modes of transport, limited number of means of transport for each mode);

- different products are combined in one batch of transport;
- the cost of supplies is presented in the form of a function (in this approach linear function of fixed and variable costs);
- different decision levels are considered simultaneously;

Decision levels in supply chains are mainly classified by the extent or effect of the decision to be made in terms of time. For instance, at the strategic level, the decisions made in relation to selecting production, storage and distribution locations, etc. should be taken. At the tactical level however, the aspects such as production and distribution planning, assigning production and transport capacities, inventories and managing safety inventories are identified. Finally, at the operational level, replenishment and delivery operations are classified (Huang et al. 2003). Most of the reviewed works focus on the tactical decision level (Perea-lopez et al. 2003; Park 2005; Rizk et al. 2006; Selim et al. 2008; Lee and Kim 2000; Chern and Hsieh 2007; Torabi and Hassini 2008). Only few works deal with the problems taken together for the different decision levels (Kanyalkar and Adil 2005; Jang et al. 2002). Therefore, the motivation behind this work is to suggest an approach to decision support in the supply chain that can provide decision support at various levels of decision-making, taking into account the many types of constraints (time, capacity, environment etc.). The proposed approach will be a demand-driven model.

Due to the size and complexity of decision problems occurring in the automotive supply chain, the proposed two-phase approach will provide a possibility of combining declarative environments (constraint logic programming) and operation research (mathematical programming).

4 The Concept of Two-Phase Approach to Decision Support in Automotive Supply Chain

The solution methods of constraint logic programming (CLP) (Apt and Wallace 2006; Sitek and Wikarek 2008) and operations research (OR) have complementary strengths. Recent research shows that these strengths can be profitably combined in hybrid approaches (Williams 2009).

Under the right conditions, one need not choose between CLP and OR, but can have the best of all two worlds. There is also the issue of problem formulation. CLP and OR have developed their own distinctive modeling styles, which poses the question of how to formulate problems that are to be solved by hybrid methods. We present an approach that allows modeling and solution of the same model of decision making in both environments independently but also to communicate between them.

The nature and complexity of decision problems occurring in the automotive supply chain, a two-phase approach to decision support was proposed. The general idea of this approach is to deploy each phase in a very flexible, need-dependent way. Each phase has a different environment for modeling and implementation—mathematical programming environment (operational research) and constraint logic programming.

Figure 2 presents the assumptions on the two-phase approach. It is a combination of two environments and the ability to use their best properties to the corresponding decision problems on the one hand. On the other hand, it provides a possibility to use only one of them depending on the needs.

If you know the average demand occurring in the supply chain, based for instance on historical data (implementation of previous deliveries) and current costs related to transportation, storage, owned production capacity, environmental fees, deadlines, etc., then to optimize the necessary means of transport, volume of distribution centers and their locations the mathematical programming environment can be used for first phase (1P).

If current demand and orders are not yet known, and there is a high probability of some additional constraints in the transport, distribution, etc., then using the results of first phase (1P) related the structure and resources (means of transport

Fig. 2 Schematic concept of two-phase approach

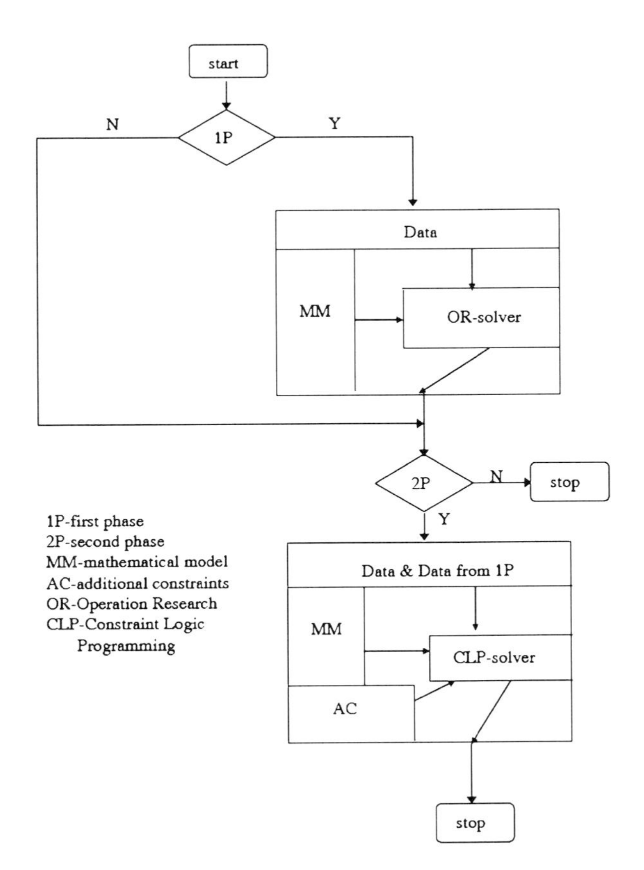

volume, distribution centers and their location etc.), we use constraint logic programming environment in second phase (2P) by introducing and optimizing individual orders based on the resources at hand.

Due to the implementation of the same model of decision making in both environments, each phase can be used individually according to the diagram in Fig. 2. For example, if the orders are known, one will only need Phase 1, but certain performance problems may occur because many decisions have to be made simultaneously concerning the chain structure, necessary resources and operational performance of the orders.

Phase 2 can be also used when the history of orders or the optimal structure are not known

4.1 Problem Formulation-Mathematical Model

In the proposed two-phase approach can implement any decision model (a mathematical model). They may be models of costs, resources and multi-criteria.

As an example in this chapter a mathematical optimization model was formulated as an integer linear programming problem (Schrijver 1998) with the minimization of costs (1) under constraints (2)–(23). Indices, parameters and decision variables in the model together with their descriptions are provided in Table 1. The proposed optimization model is a cost model that takes into account four other types of parameters, i.e., the spatial parameters (area/volume occupied by the product, distributor capacity and capacity of transport unit), time (duration of delivery and service by distributor, etc.), environment and transport The model concerns the structure of the supply shown in Fig. 3. It is a universal part of the automotive supply chain. The position of each parameter against the subsequent links of the supply chain is shown in Fig. 4. The proposed example of a mathematical model of the distribution supply chain consisting of a single period ordering. The policies of supplies such as Just In Sequence, etc. can be achieved by proper selection of orders for a given period and subsequent periods.

4.1.1 Optimization Criteria

The objective function (1) defines the aggregate costs of the entire chain and consists of five elements. The first is the fixed costs associated with the operation of the distributor involved in the delivery (e.g. distribution center, warehouse, etc.). The second part sets out the environmental costs of using various means of transport. They are dependent on the one hand the number of courses the means of transport, the other from the environmental levy, which may depend on the use of fossil fuels and carbon-dioxide emissions.

Table 1 Summary indices, parameters and decision variables of the mathematical optimization model

Symbol	Description
Indices	
k	product type (k = 1...O)
j	delivery point/customer/city (j = 1...M)
i	manufacturer/factory (i = 1...N)
s	distributor/distribution center (s = 1...E)
d	mode of transport (d = 1...L)
N	number of manufacturers/factories
M	number of delivery points/customers
E	number of distributors
O	number of product types
L	number of mode of transport
Input parameters	
F_s	the fixed cost of distributor/distribution center *s* (s = 1...E)
P_k	the area/volume occupied by product *k* (k = 1...O)
V_s	distributor *s* maximum capacity/volume (s = 1...E)
$W_{i,k}$	production capacity at factory *i* for product *k* (i = 1...N) (k = 1...O)
$C_{i,k}$	the cost of product *k* at factory *i* (i = 1...N) (k = 1...O)
$R_{s,k}$	if distributor *s* (s = 1...E) can deliver product *k* (k = 1...O) then $R_{sk} = 1$, otherwise $R_{sk} = 0$
$Tp_{s,k}$	the time needed for distributor *s* (s = 1...E) to prepare the shipment of product *k* (k = 1...O)
$Tc_{j,k}$	the cut-off time of delivery to the delivery point/customer *j* (j = 1...M) of product *k* (k = 1...O)
$Z_{j,k}$	customer demand/order *j* (j = 1...M) for product *k* (k = 1. O)
Zt_d	the number of transport units using mode of transport *d* (d = 1...L)
Pt_d	the capacity of transport unit using mode of transport *d* (d = 1...L)
$Tf_{i,s,d}$	the time of delivery from manufacturer *i* to distributor *s* using mode of transport *d* (i = 1...N) (s = 1...E) (d = 1...L)
$K1_{i,s,k,d}$	the variable cost of delivery of product *k* from manufacturer *i* to distributor *s* using mode of transport *d* (d = 1...L) (i = 1...N) (s = 1...E) (k = 1...O)
$R1_{i,s,d}$	if manufacturer *i* can deliver to distributor *s* using mode of transport *d* then $R1_{isd} = 1$, otherwise $R1_{isd} = 0$ (d = 1...L) (s = 1...E) (i = 1...N)
$A_{i,s,d}$	the fixed cost of delivery from manufacturer *i* to distributor *s* using mode of transport *d* (d = 1...L) (i = 1...N) (s = 1...E)
$Koa_{s,j,d}$	the total cost of delivery from distributor *s* to customer *j* using mode of transport *d* (d = 1...L) (s = 1...E) (j = 1...M)
$Tm_{s,j,d}$	the time of delivery from distributor *s* to customer *j* using mode of transport *d* (d = 1...L) (s = 1...E) (j = 1...M)
$K2_{s,j,k,d}$	the variable cost of delivery of product *k* from distributor *s* to customer *j* using mode of transport *d* (d = 1...L) (s = 1...E) (k = 1...O) (j = 1...M)
$R2_{sjd}$	if distributor *s* can deliver to customer *j* using mode of transport *d* then $R2_{sjd} = 1$, otherwise $R2_{s,j,d} = 0$ (d = 1...L) (s = 1...E) (j = 1...M)
$G_{s,j,d}$	the fixed cost of delivery from distributor *s* to customer *j* using mode of transport *d* (s = 1...E) (j = 1...M) (k = 1...O)

(continued)

Table 1 (continued)

Symbol	Description
$Kog_{s,j,d}$	the total cost of delivery from distributor *s* to customer *j* using mode of transport *d* (d = 1...L) (s = 1...E) (j = 1...M) (k = 1...O)
Od_d	the environmental cost of using mode of transport *d* (d = 1...L)
Decision variables	
$X_{i,s,k,d}$	delivery quantity of product *k* from manufacturer *i* to distributor *s* using mode of transport *d*
$Xa_{i,s,d}$	if delivery is from manufacturer *i* to distributor *s* using mode of transport *d* then $Xa_{i,s,d} = 1$, otherwise $Xa_{i,s,d} = 0$
$Xb_{i,s,d}$	the number of courses from manufacturer *i* to distributor *s* using mode of transport *d*
$Y_{s,j,k,d}$	delivery quantity of product *k* from distributor *s* to customer *j* using mode of transport *d*
$Ya_{s,j,d}$	if delivery is from distributor *s* to customer *j* using mode of transport *d* then $Ya_{s,j,d} = 1$, otherwise $Ya_{s,j,d} = 0$
$Yb_{s,j,d}$	the number of courses from distributor *s* to customer *j* using mode of transport *d*
Tc_s	if distributor *s* participates in deliveries, then $T_{cs} = 1$, otherwise $T_{cs} = 0$
CW	Arbitrarily large constant

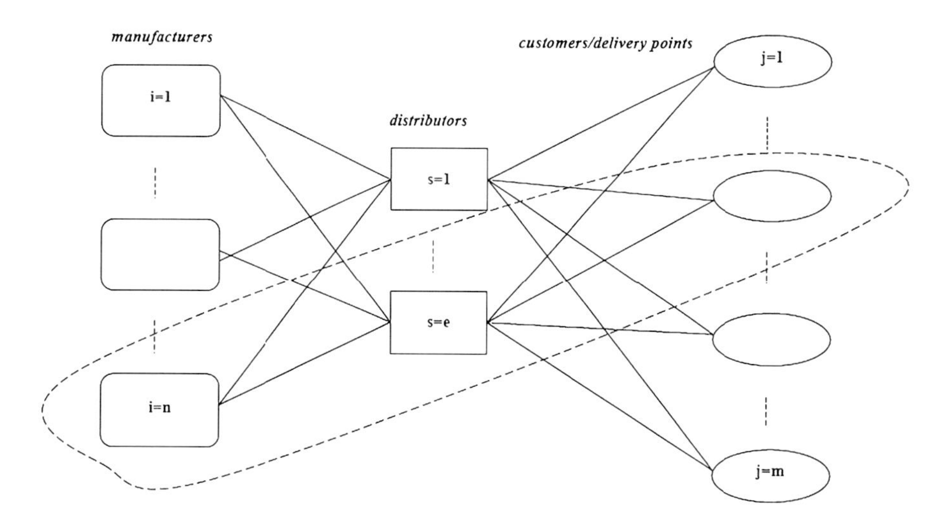

Fig. 3 The part of the supply chain network with marked indices of individual participants (elements). *Dashed line* marks one of the possible routes of delivery

The third component determines the cost of supply from the manufacturer to the distributor. Another component is responsible for the costs of supply from the distributor to the end user (the store, the individual client, etc.). The last component of the objective function determines the cost of manufacturing the product by the given manufacturer.

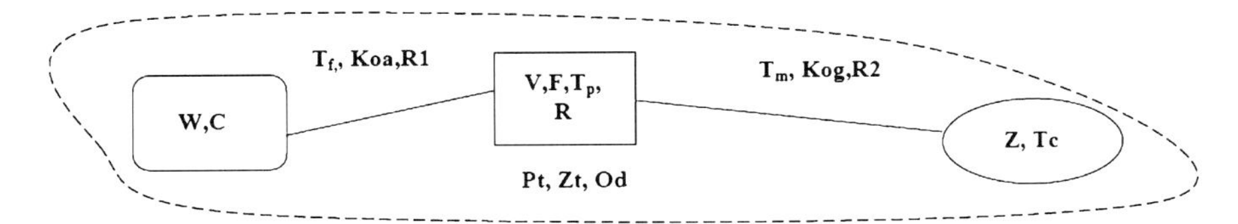

Fig. 4 The selected path of the supply chain along with the parameters that describe the individual elements and its dependencies (shared information)

$$\sum_{s=1}^{E} F_s * Tc_s + \sum_{d=1}^{L} Od_d \left(\sum_{i=1}^{N} \sum_{s=1}^{E} Xb_{i,s,d} + \sum_{s=1}^{E} \sum_{j=1}^{M} Yb_{j,s,d} \right) + \sum_{i=1}^{N} \sum_{s=1}^{E} \sum_{d=1}^{L} Koa_{i,s,d} + \sum_{s=1}^{E} \sum_{j=1}^{M} \sum_{d=l}^{L} Kog_{s,j,d} + \sum_{i=1}^{N} \sum_{k=1}^{O} \left(C_{ik} * \sum_{s=1}^{E} \sum_{d=1}^{L} X_{i,s,k,d} \right) \quad (1)$$

4.1.2 Constraints

The model was developed subject to constraints (2)–(23). Constraint (2) specifies that all deliveries of product k produced by the manufacturer i and delivered to all distributors s using mode of transport d do not exceed the manufacturer's production capacity.

$$\sum_{s=1}^{E} \sum_{d=1}^{L} X_{i,s,k,d} \leq W_{i,k} \; dla \; i = 1 \ldots N, \; K = 1 \ldots O \quad (2)$$

Constraint (3) covers all customer j demands for product k ($Z_{j,k}$) through the implementation of supply by distributors s (the values of decision variables $Y_{i,s,k,d}$). The constraint was designed to take into account the specificities of the distributors resulting from environmental or technological constraints (i.e., whether the distributor s can deliver the product k or not).

$$\sum_{s=1}^{E} \sum_{d=1}^{L} (Y_{s,j,k,d} * R_{s,k}) \geq Z_{j,k} \text{ for j} = 1 \ldots \text{M, k} = 1 \ldots \text{O} \quad (3)$$

The balance of each distributor s corresponds to constraint (4)

$$\sum_{i=1}^{N} \sum_{d=1}^{L} X_{i,s,k,d} = \sum_{j=1}^{M} \sum_{d=1}^{L} Y_{s,j,k,d} \text{ for s} = 1 \ldots \text{E, k} = 1 \ldots \text{O} \quad (4)$$

The possibility of delivery in due to its technical capabilities - in the model, in terms of volume/capacity of the distributor's is defined by constraint (5)

$$\sum_{k=1}^{O}\left(P_k * \sum_{i=1}^{N}\sum_{d=1}^{L} X_{i,s,k,d}\right) \leq Tc_s * V_s \text{ for s} = 1\ldots\text{E} \quad (5)$$

Constraint (6) ensures the fulfillment of the terms of delivery time.

$$Xa^*_{i,s,d}\mathrm{Tf}_{\mathrm{i,s,a}} + Xa_{i,s,d} * \mathrm{Tp}_{\mathrm{s,k}} + Ya^*_{s,j,d}\mathrm{Tm}_{\mathrm{s,j,d}} \leq Tc_{j,k}$$
$$\text{for i} = 1..\text{N, s} = 1\ldots\text{E, j} = 1\ldots\text{M, k} = 1\ldots\text{O, d} = 1\ldots\text{L} \quad (6)$$

Constraints (7a), (7b), (8) guarantee deliveries with available transport taken into account.

$$\mathrm{R1}_{\mathrm{i,s,d}} * \mathrm{Xb}_{\mathrm{i,s,d}} * \mathrm{Pt}_{\mathrm{d}}\mathrm{X}_{\mathrm{i,s,k,d}} * \mathrm{P}_{\mathrm{k}} \text{ for i} = 1\ldots\text{N, s} = 1\ldots\text{E, k} = 1\ldots\text{O, d} = 1\ldots\text{L} \quad (7a)$$

$$R2_{s,j,d} * Yb_{s,j,d} * Pt_d = Y_{s,j,k,d} * \mathrm{P}_{\mathrm{k}} \text{ for s} = 1\ldots\text{E, j} = 1\ldots\text{M, k} = 1\ldots\text{O, d} = 1\ldots\text{L} \quad (7b)$$

$$\sum_{i=1}^{N}\sum_{s=1}^{E} Xb_{i,s,d} + \sum_{j=1}^{M}\sum_{s=1}^{E} \mathrm{Yb}_{\mathrm{j,s,d}} \leq Zt_d \text{ for d} = 1\ldots\text{L} \quad (8)$$

Constraints (9)–(11) respectively set values of decision variables based on binary variables Tc_s, $Xa_{i,s,d}$, $Ya_{s,j,d}$.

$$\sum_{i=1}^{N}\sum_{d=1}^{L} Xb_{i,s,d} \leq CW * Tc_s \text{ for s} = 1\ldots\text{E} \quad (9)$$

$$Xb_{i,s,d} \leq CW * Xa_{i,s,d} \text{ for i} = 1\ldots\text{N, s} = 1\ldots\text{E, d} = 1\ldots\text{L} \quad (10)$$

$$Yb_{s,j,d} \leq CW * Ya_{s,j,d} \text{ for s} = 1\ldots\text{E, j} = 1\ldots\text{M, d} = 1\ldots\text{L} \quad (11)$$

Dependencies (12) and (13) represent the relationship by which total costs are calculated. In general, these may be any linear functions.

$$\mathrm{koa}_{\mathrm{i,s,d}} = \mathrm{A}_{\mathrm{i,s,d}} * \mathrm{Xb}_{\mathrm{i,s,d}} + \sum_{\mathrm{k}=1}^{\mathrm{O}} \mathrm{K1}_{\mathrm{i,s,k,d}} * \mathrm{X}_{\mathrm{i,s,k,d}} \text{ for i} = 1..\text{N, s} = 1\ldots\text{E, d} = 1\ldots\text{L} \quad (12)$$

$$\mathrm{Kog}_{\mathrm{s,j,d}} = \mathrm{G}_{\mathrm{s,j,d}} * \mathrm{Yb}_{\mathrm{j,s,d}} + \sum_{\mathrm{k}=1}^{\mathrm{O}} \mathrm{K2}_{\mathrm{s,j,k,d}} * \mathrm{Y}_{\mathrm{s,j,k,d}} \text{ for s} = 1\ldots\text{E, j} = 1\ldots\text{M, d} = 1\ldots\text{L} \quad (13)$$

The remaining constraints (14)–(23) arise from the nature of the model.

$$X_{i,s,k,d} \geq 0 \text{ for } i = 1\ldots N,\ s = 1\ldots E,\ k = 1\ldots 0,\ d = 1\ldots L \tag{14}$$

$$Xb_{i,s,d} \geq 0 \text{ for } i = 1\ldots N,\ s = 1\ldots E,\ d = 1\ldots L, \tag{15}$$

$$Yb_{s,j,d} \geq 0 \text{ for } s = 1\ldots E,\ j = 2\ldots M,\ d = 1\ldots L, \tag{16}$$

$$X_{i,s,k,d} \in C \text{ for } i = 1\ldots N,\ s = 1\ldots E,\ k = 1\ldots 0,\ d = 1\ldots L, \tag{17}$$

$$Xb_{i,s,d} \in C \text{ for } i = 1\ldots N,\ s = 1\ldots E,\ d = 1\ldots L \tag{18}$$

$$Y_{s,j,k,d} \in C \text{ for } s = 1\ldots E,\ j = 1\ldots M,\ k = 1\ldots 0,\ d = 1\ldots L \tag{19}$$

$$Yb_{s,j,d} \in C \text{ for } s = 1\ldots E,\ j = 1\ldots M,\ d = 1\ldots L, \tag{20}$$

$$Xa_{i,s,d} \in \{0,1\} \text{ for } i = 1\ldots N,\ s = 1\ldots E,\ d = 1\ldots L, \tag{21}$$

$$Ya_{s,j,d} \in \{0,1\} \text{ for } s = 1\ldots E,\ j = 2\ldots M,\ d = 1\ldots L, \tag{22}$$

$$Tc_s \in \{0,1\} \text{ for } s = 1\ldots E \tag{23}$$

4.2 Method Developed

In first phase (1P) the model was implemented in "LINGO" by LINDO Systems (www.lindo.com). "LINGO" Optimization Modeling Software is a powerful tool for building and solving mathematical optimization models. "LINGO" package provides the language to build optimization models and the editor program including all the necessary features and built-in "solvers" in a single integrated environment. "LINGO" is designed to model and solve linear, nonlinear, quadratic, integer and stochastic optimization problems. In second phase (2P) mathematical model was implemented in ECL^{i}PS^{e} (www.eclipseclp.org). ECL^{i}PS^{e} is an open-source software system for the cost-effective development and deployment of constraint programming applications, e.g. in the areas of planning, scheduling, resource allocation, timetabling, transport etc. It contains several constraint solver libraries, a high-level modeling and control language, interfaces to third-party solvers, an integrated development environment and interfaces for embedding into host environments (Schrijver 1998). ECL^{i}PS^{e} support the most common techniques used in solving Constraint (Optimization) Problems:(Constraint Programming, Mathematical Programming, Local Search). It is built around the CLP (Constraint Logic Programming) paradigm (Apt and Wallace 2006).

4.3 Computational Examples

After the implementation of LINGO (1P) and ECL^{i}PS^{e} (2P) experiments were calculated. Optimization was performed for three examples: E1, E2 and E3 (sample data based on experience gained from practical experience for the various supply chains).

All the cases relate to the supply chain with two manufacturers $(i = 1...2)$, three distributors $(s = 1...3)$, four recipients $(j = 1...4)$, four mode of transport $(d = 1...4)$ and five types of products $(k = 1...5)$. The examples differ in capacity available to the distributors (V_s) and number of transport units using mode of transport d (Z_{td}). The numeric data for all the model parameters from Table 1 are presented in Appendix A.

In the examples E1 and E2 distributors capacities are $V_1 = 2000$, $V_2 = 2500$, $V_3 = 1500$; for E3 are $V_1 = 1200$, $V_2 = 1500$, $V_3 = 1200$. Parameters Zt_d for E1 are $Zt_1 = 12$, $Zt_2 = 6$, $Zt_3 = 3$; for E2 $Zt_1 = Zt_2 = Zt_3 = 20$ and for E3 $Zt_1 = 10$, $Zt_2 = 10$, $Zt_3 = 3$. Other details are the same for all three examples. In the first stage of experiments, it was assumed that the data on customer demand (Z_{jk}) are known and fixed in a given period of optimization. Therefore, the proposed approach uses only first phase (1P). Then, the effect of parameters V_s, Zt_d on the obtained solution was investigated. The results of this phase and influence of parameters (V_s, Zt_d) are presented in the form of the objective function ($Fc^{opt} = 52105$ for E1, $Fc^{opt} = 50700$ for E2, and $Fc^{opt} = 52465$ for E3) and the optimal transportation networks.

Transportation networks diagrams showing the number of hauls (no number means one) corresponding to the optimal solutions for E1, E2, E3 are shown sequentially in Figs. 5, 6, 7.

Due to lack of space, the chapter does not contain the values of decision variables $X_{i,s,k,d}$, $Y_{s,j,k,d}$ that define the size of the orders for each customer.

In the second stage of experiments, both phases of the proposed approach were used. The experiments employed E2 data as in Appendix A. It was assumed that customer demand is not known, and in the first phase optimization (1P) was performed treating the demand as an average value for the given optimization horizon.

The result of 1P was determining the necessary resources V_s, Zt_d, W_{ik}. Table 2 compares some of these values before and after 1P. These data (received after 1P) were transferred to 2P, while other data remained the same (Appendix A).

Based on the updated data, the optimization of incoming customers' demands/orders with parameters as shown in Table 3 was performed. Orders in the second phase (2P) have priorities (Pr), which, in the simplest form, may result from the order in which the orders appear or from the order decided on by the operator. Experiments were performed for the examples E2.1, E2.2 and E2.3. They varied in terms of the number and priorities of the orders placed. In example E2.1, three orders of equal priority were introduced, five orders in example E2.2, and eights orders with assigned priorities in E2.3. The results of both phases are presented in

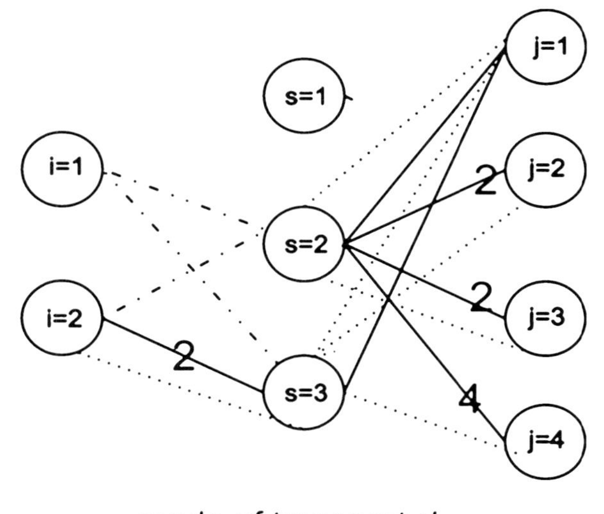

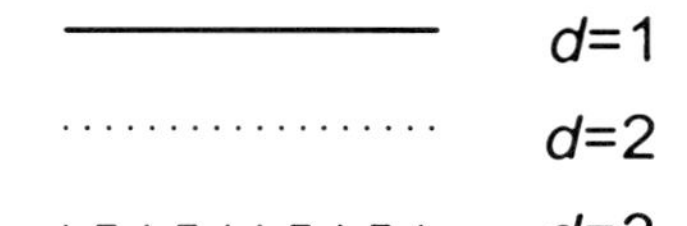

Fig. 5 Transport network of multi-modal optimal solution ($Fc^{opt} = 52105$) for E1

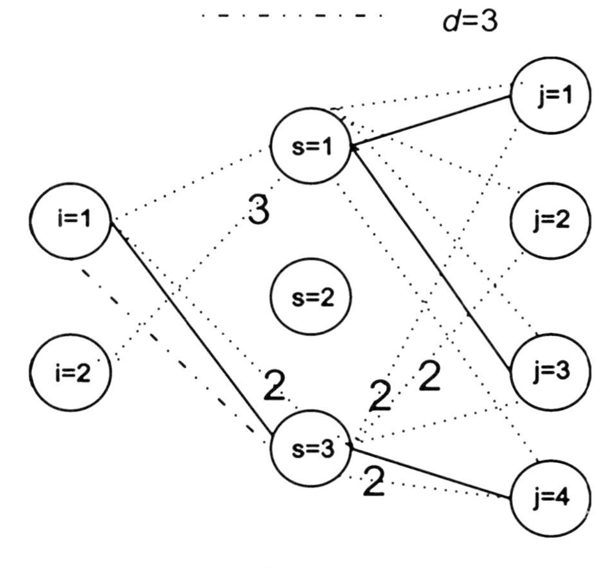

mode of transport *d*

d=1

d=2

d=3

Fig. 6 Transport network of multimodal optimal solution ($Fc^{opt} = 50700$) for E2

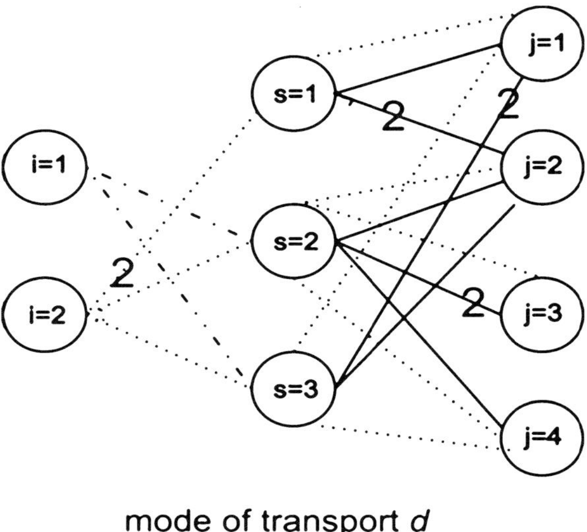

Fig. 7 Transport network of multimodal optimal solution ($Fc^{opt} = 52465$) for E3

Table 2 Values of resource parameters before and after 1P for E2

Parameter	Value before 1P	Value before 2P (after 1P)
V_1	2 000	1 375
V_2	2 500	0
V_3	1 500	1 500
Zt_1	20	4
Zt_2	20	17
Zt_3	20	1

the form of the objective function ($Fc^{opt} = 8450$ for E2.1, $Fc^{opt} = 13385$ for E2.2, and $Fc^{opt} = 15820$ for E2.3), and optimal transport network (Fig. 8, 9, 10).

5 Conclusions

The proposed two-phase approach allows a wide range of decision support in a variety of real situations that occur in the supply chain (delivery quantity of product, the number of courses, mode of transport, optimal multimodal network, used capacity, etc.). Large groups of time, resource or environmental constraints can also be included, even those emerging unexpectedly in a dynamic manner

Table 3 Incoming customers' demands/orders for E2.1, E2.2, E2.3

E2.1

j	k	Z_{jk}	Tc_{jk}	Pr
1	1	10	10	1
2	2	20	10	1
3	3	10	10	1

E2.2

j	k	Z_{jk}	Tc_{jk}	Pr
1	1	10	10	1
2	2	20	10	1
3	3	10	10	1
1	2	10	10	1
1	3	15	10	1

E2.3

j	k	Z_{jk}	Tc_{jk}	Pr	j	k	Z_{jk}	Tc_{jk}	Pr
1	1	10	10	1	3	1	10	10	2
2	1	10	10	2	3	3	10	10	2
3	3	10	10	1	3	4	10	10	3
1	2	10	10	1	4	1	10	10	3

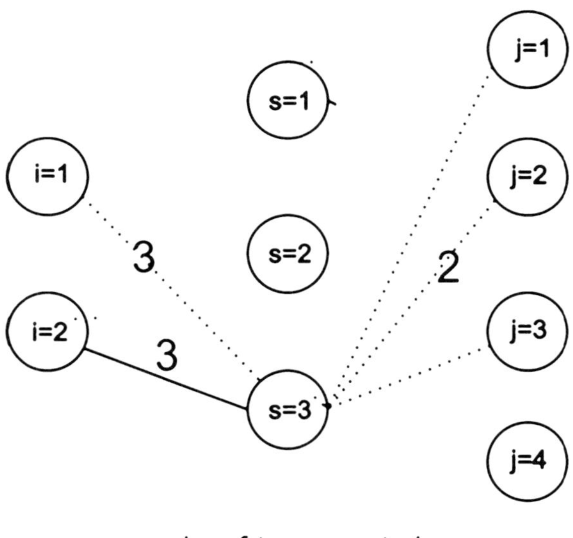

Fig. 8 Transport network of multi-modal optimal solution ($Fc^{opt} = 8450$) for E1

(declarative environment). These advantages result from the use of the possibilities and advantages of the employed modeling environments and decision-making models implementation. Mathematical programming environment (LINGO) enables efficient optimization of classical, well-defined decision problems of known structure. The constraint logic programming environment (ECL^{i}PS^e) works very effectively with decision-making problems of poor structure, descriptive,

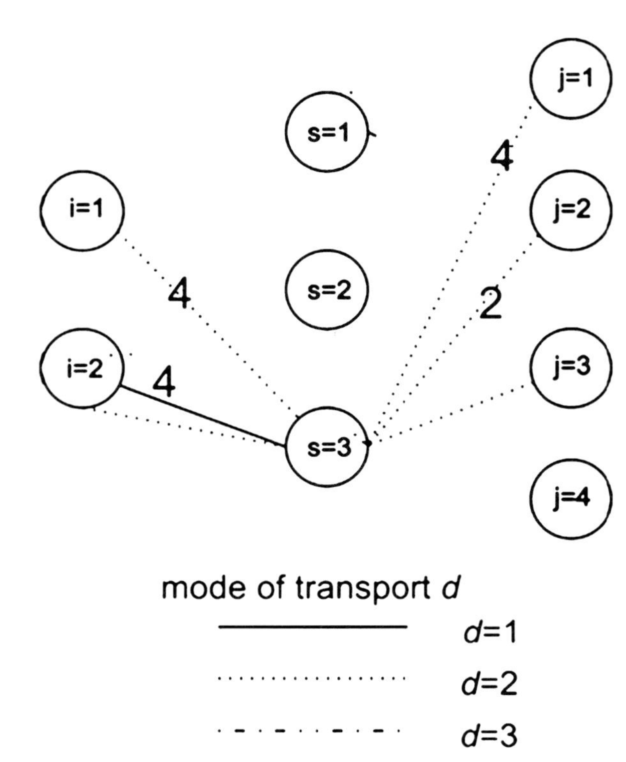

Fig. 9 Transport network of multimodal optimal solution ($Fc^{opt} = 13385$) for E2

dynamically emerging new constraints or when feasible solutions need to be quickly found. The presented approach is a framework to implement various models of decision-making. The model optimization of supply chain costs takes into account, among others multimodal transport and environmental conditions is only an example of the possibilities of using the proposed approach.

A.1 6 Appendix A

```
!P - the area/volume occupied by product k;
10 15 15 10 20~
!F - the fixed cost of distributor/distribution center s;
1200 1500 1000~
!V - distributor s maximum capacity/volume for E3 1200 1500
1200;
2000 2500 1500~
!Pt - the capacity of transport unit using mode of transport d;
```

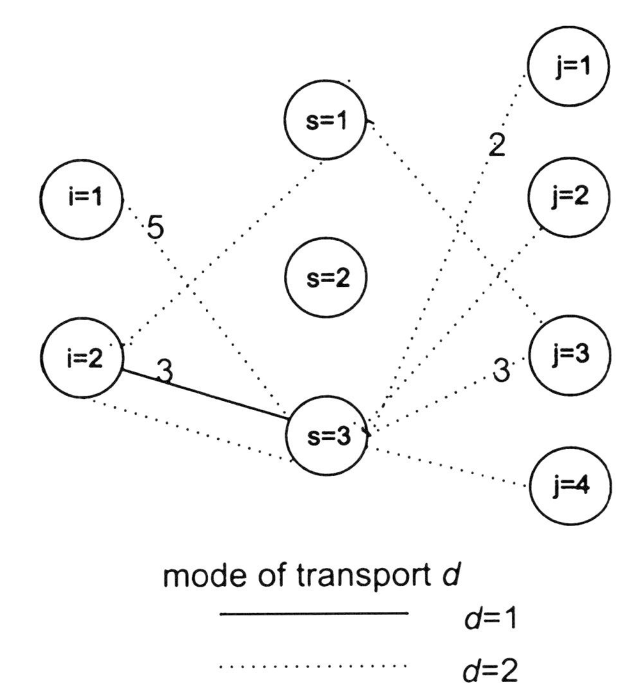

Fig. 10 Transport network of multimodal optimal solution ($Fc^{opt} = 15820$) for E3

```
60 180 600~
!Zt - the number of transport units using mode of transport
d;
!E2 20 20 20;
!E3
12 6 3 ~
!Z - customer j demand/order for product k;
10 10 15 20 15 10 0 10 10 15 10 20 0 20 0 10 0 10 0 20~
!Tcm - the cut-off time of delivery to the customer j of
product k;
10 10 10 10 20 10 10 10 10 20 10 10 10 10 20 10 10 10 10 20~
!C - the cost of product k at factory i;
100 200 200 300 300 150 210 150 250 350~
!w - production capacity at factory i for product k;
100 100 100 100 100 100 100 100 100 100~
!R - if distributor s can deliver product k;
1 1 1 1 0 1 1 0 1 1 1 1 1 0 1~
!Tp - the time needed for distributor;
!to prepare the shipment of product k;
```

```
2 2 2 2 2 1 1 1 1 1 3 3 3 3 3~
!A - the fixed cost of delivery from manufacturer i;
!to distributor s using mode of transport d;
10 20 40 12 24 42 5 10 25 5 10 20 10 20 40 15 25 35 ~
!R1 - if manufacturer i can deliver to distributor s;
!using mode of transport d;
1 1 1 1 1 1 1 1 1 1 1 0 1 1 1 1 1 0 ~
!Tf - the time of delivery from manufacturer i to distributor
s;
!using mode of transport d;
2 3 4 1 2 3 1 2 3 4 6 7 4 6 7 4 6 7 ~
!G - the fixed cost of delivery from distributor s;
!to customer j using mode of transport d;
2 4 10 2 5 12 14 12 20 15 13 30 4 8 16 3 6 15 5 10 15 2 4 10
2 4 11 3 6 14 6 10 20 4 8 20 ~
!R2 - if distributor s can deliver to customer j;
!using mode of transport d;
1 1 0 1 1 1 1 1 1 1 1 1 1 1 1 1 1 1 1 0 1 1 0 1 1 0 1 1 1
1 1 1 1 1 1 ~
!Tm - the time of delivery from distributor s;
!to customer j using mode of transport d;
1 1 2 1 1 2 1 1 2 1 1 2 1 1 2 1 1 2 1 1 2 1 1 2 1 1 2 1 1
2 1 1 2 1 1 2 ~
!K1 - the variable cost of delivery of product k;
!from manufacturer i to distributor s using mode of trans-
port d;
3 2 1 3 2 1 3 2 1 3 2 1 3 2 1 4 2 1 4 2 1 4 2 1 4 2 1 4 2 1
3 2 1 3 2 1 3 2 1 3 2 1 3 2 1 4 2 1 4 2 1 4 2 1 4 2 1 4 2 1
3 2 1 3 2 1 3 2 1 3 2 1 3 2 1 4 2 1 4 2 1 4 2 1 4 2 1 4 2 1 ~
!K2 - the variable cost of delivery of product k;
!from manufacturer i to distributor s using mode of trans-
port d;
3 2 1 3 2 1 3 2 1 3 2 1 3 2 1 3 2 1 3 2 1 3 2 1 3 2 1 3 2 1
3 2 1 3 2 1 3 2 1 3 2 1 3 2 1 3 2 1 3 2 1 3 2 1 3 2 1 3 2 1
3 2 1 3 2 1 3 2 1 3 2 1 3 2 1 3 2 1 3 2 1 3 2 1 3 2 1 3 2 1
3 2 1 3 2 1 3 2 1 3 2 1 3 2 1 3 2 1 3 2 1 3 2 1 3 2 1 3 2 1
3 2 1 3 2 1 3 2 1 3 2 1 3 2 1 3 2 1 3 2 1 3 2 1 3 2 1 3 2 1
3 2 1 3 2 1 3 2 1 3 2 1 3 2 1 3 2 1 3 2 1 3 2 1 3 2 1 3 2 1 ~
!Od -- the environmental cost of using mode of transport d;
10 30 400 ~
!CW - Arbitrarily large constant;
10000~
```

References

Apt K, Wallace M (2006) Constraint logic programming using eclipse, Cambridge University Press, Cambridge
Beamon BM, Chen VCP (2001) Performance analysis of conjoined supply chains. Int J Prod Res 39:3195–3218
Chern CC, Hsieh JS (2007) A heuristic algorithm for master planning that satisfies multiple objectives. Comput Oper Res 34:3491–3513
Dietz F, Lang NS, Maurer A (2004) Beyond cost reduction: reinventing the automotive oem-supplier interface, The Boston Consulting Group, Boston
Drucker P (2003) The essential drucker: the best of sixty years of peter drucker's essential writings on management. Harper Business, Harper Collins, New York
Heywood J (2006) Fueling Our Transportation Future. Sci Am 295(3):60–63
Huang GQ, Lau JSK, Mak KL (2003) The impacts of sharing production information on supply chain dynamics: a review of the literature. Int J Prod Res 41:1483–1517
Hugo WMJ, Badenhorst-Weiss JA, Van Biljon EHB (2004) Supply chain management: logistics in perspective, 3rd edn. Van Schaik, Pretoria
Jang YJ, Jang SY, Chang BM, Park J (2002) A combined model of network design and production/distribution planning for a supply network. Comput Ind Eng 43:263–281
Jung H, Jeong B, Lee CG (2008) An order quantity negotiation model for distributor-driven supply chains. Int J Prod Econ 111:147–158
Kanyalkar AP, Adil GK (2005) An integrated aggregate and detailed planning in a multisite production environment using linear programming. Int J Prod Res 43:4431–4454
Kelly T (2012) The automotive supply chain in the new normal: analysis of the industry and its supply chain opportunities. JDA Software Group, Inc, Ontario Institute of PMAC
Lee YH, Kim SH (2000) Optimal production–distribution planning in supply chain management using a hybrid simulation-analytic approach. In: Proceedings of the 2000 winter simulation conference 1 and 2, pp 1252–1259
Park YB (2005) An integrated approach for production and distribution planning in supply chain management. Int J Prod Res 43:1205–1224
Perea-lopez E, Ydstie BE, Grossmann IE (2003) A model predictive control strategy for supply chain optimization. Comput Chem Eng 27:1201–1218
Rizk N, Martel A, D'amours S (2006) Multi-item dynamic production–distribution planning in process industries with divergent finishing stages. Comput Oper Res 33:3600–3623
Schrijver A (1998) Theory of linear and integer programming. John Wiley & Sons, Chichester. ISBN 0-471-98232-6
Selim H, Am C, Ozkarahan I (2008) Collaborative production–distribution planning in supply chain: a fuzzy goal programming approach. Transp Res Part E-Logistics Transp Rev 44:396–419
Shapiro JF (2001) Modeling the supply chain. Duxbury Press, New York. ISBN 978-0-534-37741
Simchi-Levi D, Kaminsky P, Simchi-Levi E (2003) Designing and managing the supply chain: concepts, strategies, and case studies. McGraw-Hill, New York. ISBN 978-0-07-119896-7
Sitek P, Wikarek J (2008) In: Nguyen NT et al (eds) A Declarative Framework for Constrained Search Problems. New Frontiers in Applied Artificial Intelligence, Lecture Notes in Artificial Intelligence, vol 5027. Springer-Verlag, Berlin-Heidelberg: 728–737
Tang D, Qian X (2008) Product lifecycle management for automotive development focus-ing on supplier integration. Comput Ind 59:288–295
Timpe CH, Kallrath J (2000) Optimal planning in large multi-site production networks. Eur J Oper Res 126:422–435
Torabi SA, Hassini E (2008) An interactive possibilistic programming approach for multiple objective supply chain master planning. Fuzzy Sets Syst 159:193–214
Venkat K (2007) Analyzing and optimizing the environmental performance of supply chains. In: Proceedings from the ACEEE summer studies on energy efficiency in industry

Venkat K, Wakeland W (2006). Is lean necessarily green? In: Proceedings of the 50th annual meeting of the ISSS. International Society for the Systems Sciences, York
Williams HP (2009) Logic and Integer Programming. Springer, Berlin
www.lindo.com
www.eclipseclp.org

The Design and the Improvement of Reverse Logistics for Discarded Tires in Japan

Kuninori Suzuki and Nobunori Aiura

Abstract In this study, which focuses on discarded tires, the possibility of constructing a reverse logistics network over a wide area and the integration or aggregation of logistical bases is examined. This simulation employs cluster-first/route-second method, and local search. This procedure consists of three stages. The first is to analyze how to collect discarded tires and how to transport them to the factories as thermal fuels. The second portion is to consider how to improve the actual reverse logistics system for discarded tires with the numerical results of the simulation. The third is the design and the improvement of the reverse logistics system with the suggestion of the cost-reduction measures.

Keywords Reverse logistics · Thermal recycle · Forward logistics · Collection and transport · Discarded tires

1 Introduction

This study aims to analyze the present situation for collection of discarded tires in a series of flows in a reverse logistics network and to simulate cooperative reverse logistics.

In Japan, many used automobile tires, i.e., discarded tires, are collected from gas stations, tire dealers, and wreckers. Company A, which has provided various data for use in this study for the simulation, performs intermediate treatment of

K. Suzuki (✉)
Bunka Fashion Graduate University, 3-22-1, Yoyogi, Shibuya-ku, Tokyo, Japan
e-mail: ku-suzuki@bunka.ac.jp

N. Aiura
Hokkai School of Commerce, 6-6-10, Toyohira, Toyohira-ku, Sapporo, Japan
e-mail: aiura@hokkai.ac.jp

P. Golinska (ed.), *Environmental Issues in Automotive Industry*,
EcoProduction, DOI: 10.1007/978-3-642-23837-6_8,

collected discarded tires and recycles them as fuel chips in its own factories, and then distributes the fuel chips to paper and steel mills. This study reproduces this reverse logistics network for discarded tires and investigates the present situation and problems.

The studies on forward logistics, including development, production, sales, and consumption of products, are now becoming remarkably advanced.

On the other hand, studies on reverse logistics, including collection, transportation, intermediate treatment, recycling, and reuse of wastes, have not attracted as much attention from researchers.

However, nowadays, the companies are required to have responsibilities for their industrial wastes, and they need deal with it effectively. The studies on reverse logistics for the discarded tires, for example, are more and more important for the industrial world.

In 1994, Japan Automobile Tire Manufacturers Association (JATMA) unified the names of discarded tires into "discarded tires." Since fuel chips are valuable materials, their transportation and distribution belong to forward logistics instead of the reverse logistics performed by collection and transportation enterprises.

Many discarded tires in Japan weren't used any more after they are changed at the gas stations, car shops and scrap processors, but currently many plants need fuel chips for their boilers. There are 5 types of treated chips produced from discarded tires, because chips used for thermal recycling in paper and steel mills differ according to the size of the boiler's input port. The 5 types are: whole tire, 32-cut chip (obtained by cutting whole tires into 32 pieces), 16-cut chip (obtained by cutting whole tires into 16 pieces), 4-inch chip, and 2-inch chip. The Samukawa Factory produces rubber hydrocarbon and liquid rubber in addition to these 5 types of chips. Rubber hydrocarbon and liquid rubber are generated with a thermal decomposer that uses whole tires, cut tires, special whole tires (used for forklifts, bicycles, etc.), wire inserted tires, caterpillars, and rubber products in industrial wastes.

Compared to paper sledge, wood chips, used in thermal recycling as well, the fuel chips of discarded tires are used preferably from cement factories, paper and steel mills. Especially the boilers in the paper mills which have increased the demand for the fuel chips produced by recycling discarded tires, as recently oil prices have continued to be raised. The valid use of discarded tires have been established rather than illegally dumped in need. The more the efficiency of its reverse logistics system is promoted, the stronger the demand will be stronger.

That is to say, the discarded tires are actually valuable as fuel chips, and it is necessary to build the effective recycle system.

This study considers collection and transportation, reprocessing, and recycling of discarded tires and the distribution of recycled products. It simulates the actual situation of the reverse logistics system, and investigates cost-reduction measures to realize efficient reverse logistics systems.

2 Definition and Range of Reverse Logistics

2.1 Reverse Logistics in Previous Studies

Firstly to clarify the definition and range of reverse logistics is necessary. As shown in Fig. 1, the range of reverse logistics is defined as a series of processes, including collection and transportation, reprocessing, recycle/reuse and final disposal of products after leaving the site of waste generation, and the reverse logistics system has links with the forward logistic system.

The Reverse logistics system should be considered not only collection and transportation, but also the linkage with the forward logistics, which means how to make recycled products efficiently back to the market. The border line between the forward logistics and the reverse logistics is not clear. In other words, the range indicates a series of flows, including collection and transportation, intermediate treatment, and final disposal of wastes, and the ultimate distribution of recycled products to markets.

To clarify its image, Fleischmann and Bloomhof-Ruwaard (2004) pointed out that linking reverse logistics to a recycling system creates difficulty in defining the range of reverse logistics, as shown in Fig. 2. The overall logistics task consists of a series of reverse logistics and forward logistics. In particular, when also taking into account the reprocessing, recycle and reuse after collection and transportation, the range of the forward logistics is considered to have a broad overlap with that of the reverse logistics. Therefore, the effective reverse logistics system should be examined with this overall logistics task as in Fig. 2.

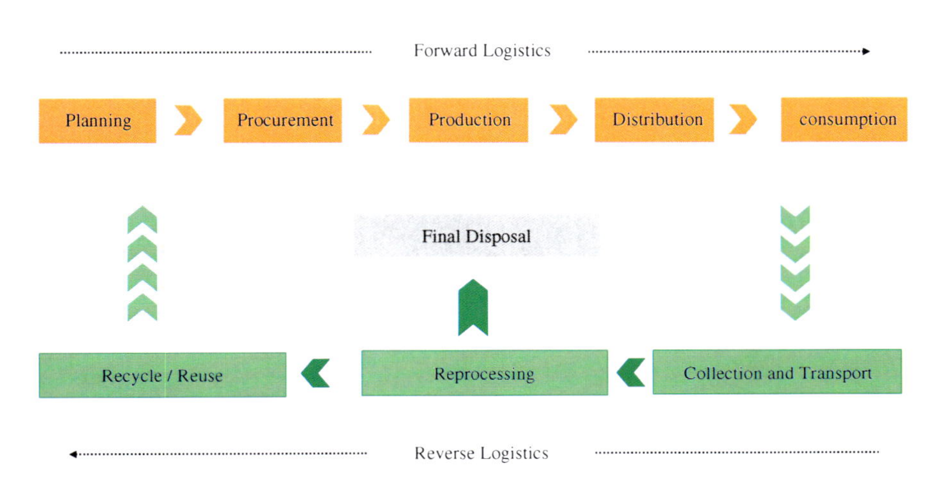

Fig. 1 Range of reverse logistics (Suzuki 2010)

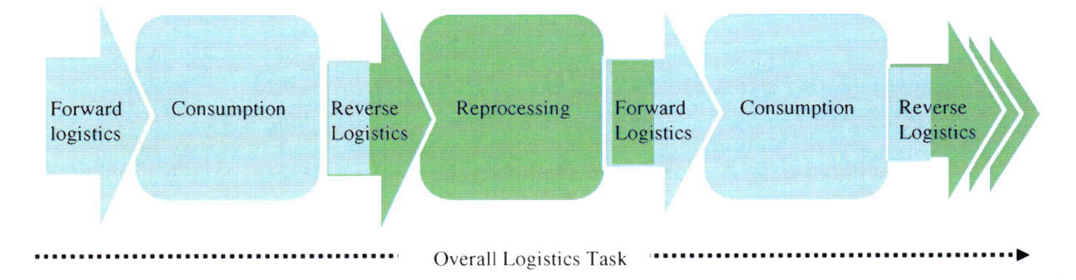

Fig. 2 Reverse logistics in the overall logistics task

2.2 Range of Reverse Logistics of Discarded Tires

The reuse of automobile tires (i.e., for discarded tires to be returned to tire markets as retreaded tires after a recycling process) is currently very low. However, after collection from gas stations and wreckers, many discarded tires are given an intermediate treatment for conversion to fuel chips, and the fuel chips are used in boilers in cement factories, and paper and steel mills as a form of thermal recycling. However, as the range of the forward supply chain of products is considered broadly overlap that of the reverse chain of wastes, as mentioned above, this study adopts the idea that the transportation of fuel chips belongs to reverse logistics.

As forward logistics, the company transports and distributes only products from discarded tires, which have undergone intermediate treatment in its own factory for thermal recycling. Figure 3 shows a material flow of discarded tires. When

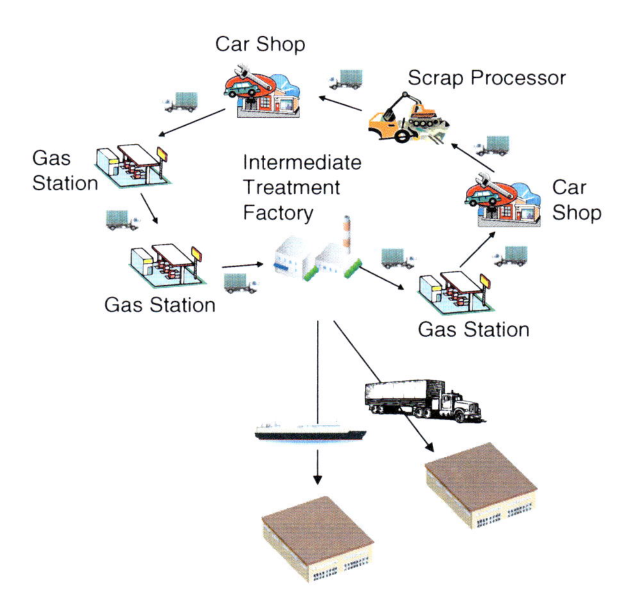

Fig. 3 Material flow of discarded tires

enterprises such as gas stations, car shops are sources of discarded tires, the discarded tires are treated as industrial wastes.

When collecting discarded tires, a person in charge of discarded tires in the source enterprise is obliged to be present at the collection point and the collection is performed using a manifesto. Therefore, the working time at the collection points, such as retailing stores, required for collecting discarded tires in reverse logistics tends to be longer than generally expected.

In the case where the same enterprise performs collection, transportation, and intermediate treatment of discarded tires, as well as distribution of fuel chips derived from discarded tires to cement factories, paper and steel mills as thermal recycling, the enterprise must be engaged in the area of reverse logistics as well as of forward logistics, including storage and distribution of the recycled products.

3 The Actual State of the Company A

Company A, which recycles discarded tires at Serizawa, Kanagawa Prefecture, has licenses for collection, transportation, and intermediate treatment of discarded tires, and for transportation of fuel chips in forward logistics.

The company recycles discarded tires into fuel chips, which it then sells to major cement factories, paper and steel mills. Discarded tires are collected from gas stations, scrap processors, and car shops in the Kanto district, or metropolitan area, and then converted into fuel chips in the company's intermediate treatment factory. The fuel chips are then directly transported to cement factories, paper and steel mills using 10-ton trucks or to steel mills using ships from the Kawasaki sea port.

A 4-ton/10-ton truck collects discarded tires from gas stations, car shops, and scrap processors from Monday through Saturday. Since these tires cannot be collected all at once along the truck's collection route, they are collected two or three times a day.

Company A's business scheme can be divided into two systems (i.e., a collection system for discarded tires and a delivery system for fuel chips after collection and intermediate treatment of discarded tires for thermal recycling).

4 Simulation Analysis of the Collection System for Discarded Tires

4.1 Purpose of Simulation Analysis

Reproducing the actual state of the company A's collection system for discarded tires and performing a scenario analysis of the future possibility of the collection system allows investigation of the present situation. It also allows exploration of further possibilities for the collection system for discarded tires and the measures needed to realize an efficient reverse logistics system.

4.2 Outline of the Simulation Model

As Fig. 4 shows, the simulation model used in this study is composed of three echelons (Smichi-Levi et al. 2002): the collection complex, the transshipment and storage complex, and the intermediate treatment factory.

Based on the management situation of company A's collection system for discarded tires and conditions set for the scenario analysis, the following three desirable situations are investigated from the viewpoints of the environment, load, and cost of the entire collection system, using a simulation model:

- Desirable situations of complexes (nodes) belonging to each echelon, i.e., how often they collect by truck, what delivery route they use by truck in Fig. 4a

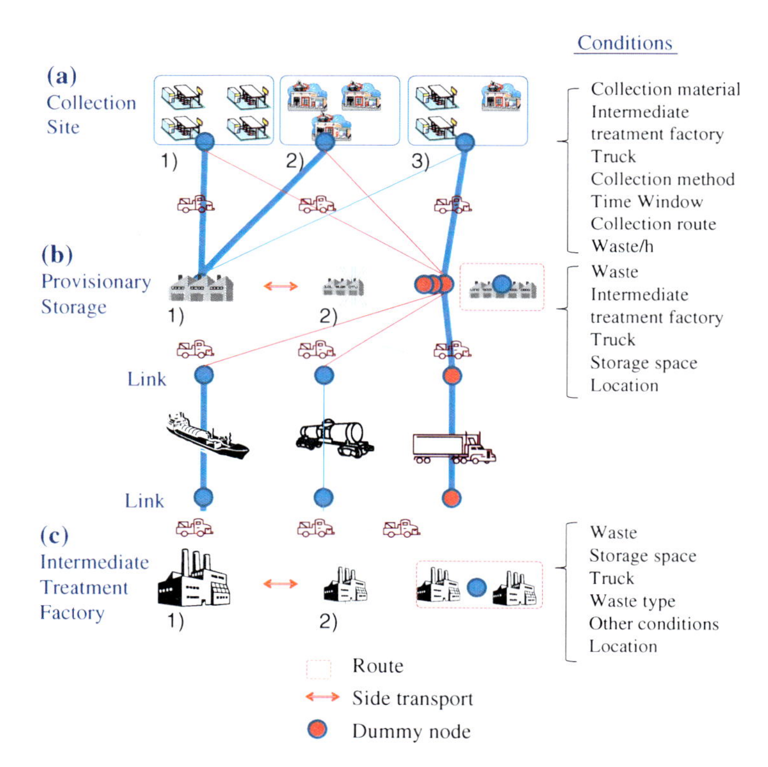

Fig. 4 Setting of echelons for discarded tires collection system

- Desirable situation of movement (links) between complexes belonging to the same echelon, i.e., the side transport between the depots in Fig. 4b provisionary storage, and Fig. 4c intermediate treatment factory
- Desirable situation of movement (links) between complexes depending to different echelons, i.e., transportation mode between a depot and an intermediate treatment factory

The actual investigation is performed as described below. Collection complexes belonging to echelons located at the upper part of Fig. 4 (such as gas stations, car shops and auto accessory stores) are used as subjects, and the study investigates a collection device at each complex in the same echelon, a collection route between collection complexes, and routes to different echelons (such as transshipment and storage complexes and intermediate treatment factories).

As explained above, based on the scenarios for the simulations, such as the expansion of the collection area, the placement of the depot, the introduction of detachable small container which may reduce truck drivers' working time at the collection sites, the costs are calculated. In addition, the scenarios depend on the node locations in collection sites, provisionary storages, and intermediate treatment factories, and also the links between the nodes over echelons.

For example, in Fig. 4, the discarded tires collected at Collection Site (a) and (b) via Provisionary Storage site 1 transport to Intermediate Treatment Factory (1) by sea. On the other hand, the discarded tires collected in the Collection Site (c) transport to Intermediate Treatment Factory (1) by truck.

When the entire area of Chiba Prefecture is assumed to be the future collection area, establishing a transshipment and storage complex (temporary storage complex) in Chiba Prefecture and performing batch transportation between the temporary storage complex and an intermediate treatment factory in Kanagawa Prefecture may be more efficient than consolidating collection complexes into one or two in Kanagawa Prefecture and performing real-time collection.

4.3 Algorithm and Preconditions for the Simulation

A simulation model used for the reverse logistics system of discarded tires consists mainly of routes between collection complexes. An algorithm using the method of cluster-first/route-second method (Kubo et al. 2001) and local search (Toth and Tramontani 2008) for allocating vehicles in forward logistics is based on an algorithm suitable for the collection system. The method to determine an efficient route first extracts a route, which can be expected to be the geometrically shortest, from Hamiltonian paths connecting collection sites to be travelled. Subsequently, the method improves the extracted route by a local search.

Provided a reference point and a baseline, the travelling path, as the initial solution, with a small angle in ascending order between the line connecting collection sites and the baseline to the reference point can be found.

The collection plan is established based on the following 6 steps under the scenarios which are explained earlier:

Step 1 Select a store (collection point) such as gas station, which has the most remarkable problem in allocating vehicles, such as geographically distant location

Step 2 Evaluate the existence of an efficient route between collection points, which satisfies the constraints, after adding a collection point nearest to the selected collection point

Step 3 Confirm that the predetermined conditions are satisfied after adding the second-nearest point, when the predetermined conditions have been satisfied in Step 2

Step 4 Evaluate the second-nearest point after excluding the nearest point to the selected collection point, when the predetermined conditions have not been satisfied in Step 2

Step 5 Determine a route for the selected store after performing Steps 1–4 on all stores

Step 6 Determine routes for remaining stores by performing Steps 1–4, after determining the route in Step 5

4.4 Outline of Information on Collection Points

At present, company A has 150 collection points for discarded tires in the Kanto district, centering on Tokyo Metropolis and Kanagawa Prefecture in Fig. 5. In the future, the company will introduce a cooperative collection system and extend the collection area for discarded tires, concentrating on Chiba Prefecture.

Assuming an increase in the number of collection points to 520, data on potential collection points are prepared in addition to data on existing collection points as shown in Fig. 6. A simulation is performed based on consolidation of complexes and information on collection points in the company A's collection area, to be incorporated into the cooperative collection system, and the effect of the increase in the number of collection points on the operational efficiency is examined.

The cooperative collection system can reduce not only the number of vehicles, and their total travelling distances, but also the facility expense, personnel expense, and other fixed expenses.

4.5 Numerical Results

The numerical results of the simulations are shown in Tables 1, 2, 3, and 4.

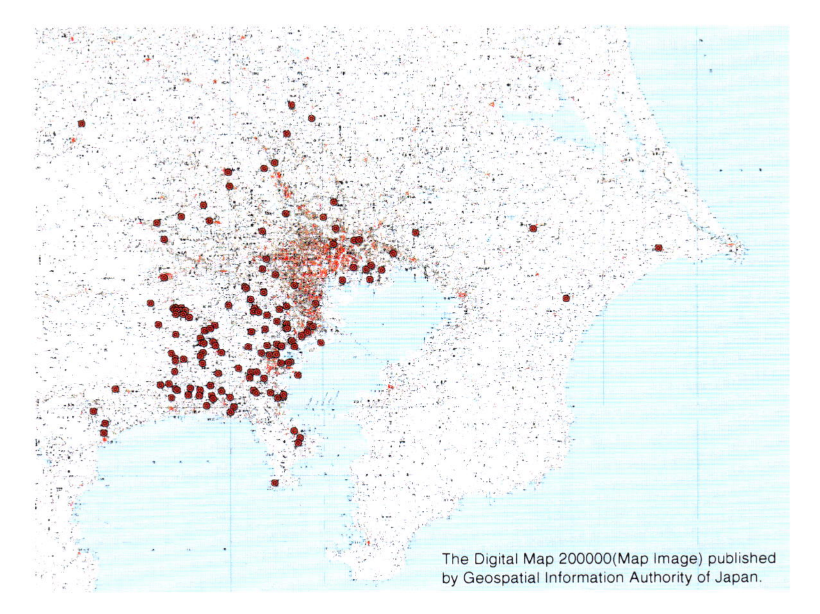

Fig. 5 Actual collection site

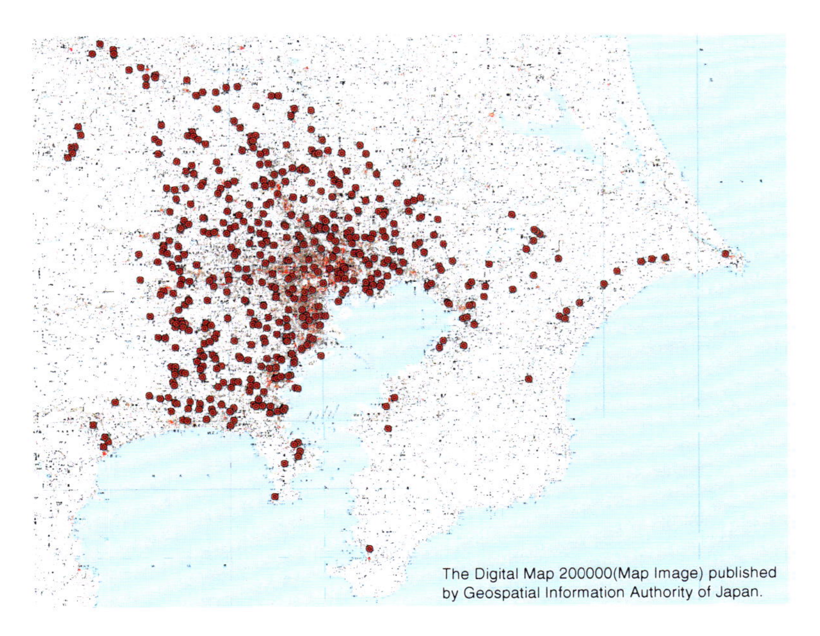

Fig. 6 Potential collection site

Table 1 Mileage result of the actual network

Date	Truck type (ton)	Driving mileage (km)	Driving time (h)	Working time (at collection sites) (h)	The number of discarded tires	The number of trucks
DATE:0	4	1,499	50	45	1,999	11
DATE:1	4	1,212	40	44	1,624	10
DATE:2	4	1,203	40	45	2,048	9
DATE:3	4	1,499	50	45	1,999	11
DATE:4	4	1,212	40	44	1,624	10
DATE:5	4	1,203	40	45	2,048	9
DATE:0	10	1,057	35	8	1,638	5
DATE:1	10	824	27	10	1,220	5
DATE:2	10	802	27	8	1,298	5
DATE:3	10	1,057	36	8	1,638	5
DATE:4	10	824	27	10	1,220	5
DATE:5	10	802	27	8	1,298	5
	Total	13,193	440	321	19,654	90

Table 2 Cost result of the actual network

Date	Truck type (ton)	Variable expense (driving) (yen)	Variable expense (at collection site) (yen)	Personnel expense (yen)	Fixed expense (yen)	Total (yen)
DATE:0	4	22,705	6,086	77,400	33,000	138,191
DATE:1	4	18,352	5,987	74,000	30,000	128,339
DATE:2	4	18,221	6,094	69,600	27,000	120,915
DATE:3	4	22,705	6,086	77,400	33,000	139,191
DATE:4	4	18,352	5,987	74,000	30,000	128,339
DATE:5	4	18,221	6,094	69,600	27,000	120,915
DATE:0	10	22,027	1,560	46,250	25,000	94,837
DATE:1	10	17,165	1,917	41,250	25,000	85,332
DATE:2	10	16,695	1,566	41,250	25,000	84,511
DATE:3	10	22,027	1,560	46,250	25,000	94,837
DATE:4	10	17,165	1,917	41,250	25,000	85,332
DATE:5	10	16,695	1,566	41,250	25,000	84,511
	Total	230,330	46,420	699,500	330,000	1,306,250

Currently, collection routes are decided by the day of the week. To grasp the actual situation, the program was run using data for 150 sites, utilizing 4-ton and 10-ton trucks.

Table 1 shows the numerical results of truck types, driving mileage, driving time, working time (at collection sites), the number of discarded tires, and the number of trucks. "DATE" means the day of the week, i.e., DATE: 0 means Monday, and DATE: 5 means Saturday. Table 2 shows the cost results by the week, the personnel expenses, the variable expenses, the fixed expenses, and the total cost. The personnel expenses are the truck drivers' wages; variable expenses

Table 3 Mileage result of the potential network

Date	Truck type (ton)	Driving mileage (km)	Driving time (h)	Working time (at collection sites) (h)	The number of discarded tires	The number of trucks
DATE:0	4	8,402	535	371	9,965	107
DATE:1	4	9,625	621	359	10,901	120
DATE:2	4	10,060	647	338	11,213	122
DATE:3	4	8,402	535	371	9,965	107
DATE:4	4	9,625	621	359	10,901	120
DATE:5	4	10,060	647	338	11,213	122
DATE:0	10	620	41	14	3,600	10
DATE:1	10	1,206	72	17	4,800	14
DATE:2	10	1,388	75	14	3,200	14
DATE:3	10	620	41	14	3,600	10
DATE:4	10	1,206	72	17	4,800	14
DATE:5	10	1,388	75	14	3,200	14
	Total	62,602	2,087	1,153	45,620	406

Table 4 Cost result of the potential network

Date	Truk type (ton)	Variable expense (driving) (yen)	Variable expense (at collection site) (yen)	Personnel expense (yen)	Fixed expense (yen)	Total (yen)
DATE:0	4	127,270	26,182	401,400	168,000	722,852
DATE:1	4	145,803	25,389	435,800	186,000	792,992
DATE:2	4	152,394	23,535	441,200	189,000	806,129
DATE:3	4	127,270	26,182	401,400	168,000	722,852
DATE:4	4	145,803	25,389	435,800	186,000	792,992
DATE:5	4	152,394	23,535	441,200	189,000	806,129
DATE:0	10	12,913	1,325	41,250	25,000	80,488
DATE:1	10	25,127	1,757	69,000	40,000	135,884
DATE:2	10	28,914	1,586	74,500	45,000	150,000
DATE:3	10	12,913	1,325	41,250	25,000	80,488
DATE:4	10	25,127	1,757	69,000	40,000	135,884
DATE:5	10	28,914	1,586	74,500	45,000	150,000
	Total	984,842	159,548	2,926,300	1,306,000	5,376,690

are fuel expenses; and the fixed expenses include of insurance and vehicle expenses.

Tables 3 and 4 show the results for the potential network as explained in Fig. 6. The scale merit and cost merit cannot be determined with the increase of the number of collection sites. The mileage distance increases, compared to the number of the trucks and collection lots of discarded tires.

The number of collection points that a truck was able visit was restricted by the travel time required and the number of discarded tires collected. In the case where there were a fixed number of collections, the number of vehicles used was not affected by the number of collection points.

4.6 Cost-Reduction Measures

As the results in Tables 3 and 4 show, the cost merit of the increase in the number of collection points cannot be determined. That is to say, other cost-reduction measures should be considered. It is possible to reduce the working time (at collection sites) shown in Tables 1 and 3.

The collection of discarded tires requires a lot of time at each site. However, the effective measures, such as collection infrastructure (Beullens et al. 2004), can reduce its time and costs. The possible measures include:

- Introduction of an RFID system for the inspection of discarded tires at the collection sites
- Introduction of an Electronic manifesto system and an ASN system (Advanced shipment/Collection notice system) for efficient procedures of collecting discarded tires at the collection sites
- Introduction of removable containers for reducing the loading time and waiting time at collection sites

Based on these cost-reduction measures, the scenarios for the simulations were constructed. The numerical results are shown in Table 5. Compared to the actual network, the costs for each scenario can be reduced. A "full complement" solution includes an RFID system, an electronic manifesto system and an ASN system, and removable containers.

Concerning removable containers (10 10-ton containers and 25 4-ton containers) for collecting discarded tires, it has been already introduced at some collection points.

However, in the case of the container collecting operation, a vehicle goes to only one collection point per operation. The number of vehicles could be reduced by using them in a rotation schedule. Since the container collection operation took time, one vehicle was planned to perform 2.5 container collections per day.

Table 5 Cost result of the measures

Cost reduction measure	Total cost (yen)
Actual state	1,306,250
RFID system	1,221,332
Electronic Manifesto and ASN	1,128,018
RFID system, electronic manifesto, and ASN	1,087,960
Removable containers	1,021,432
Full complement	928,032

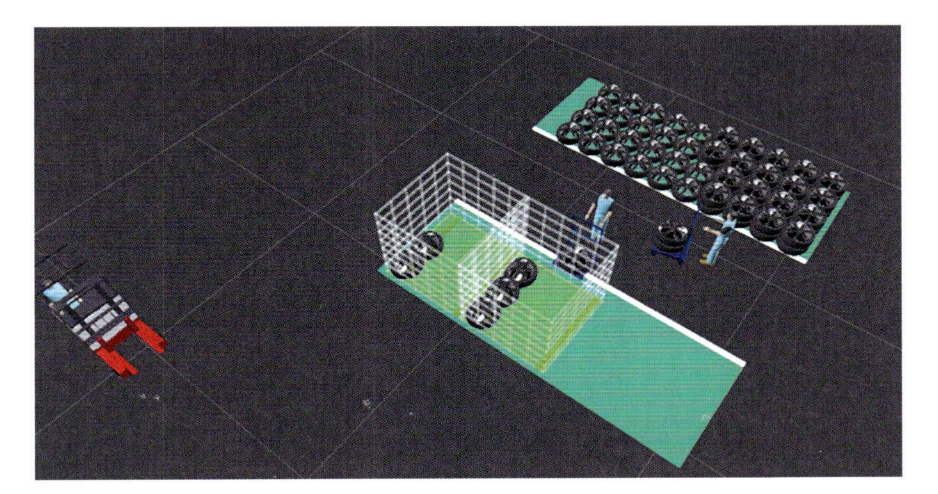

Fig. 7 New type of removable container

In general, a vehicle could perform 2–2.5 container collection operations per day. The number of vehicles was determined based on the number of collection points. In the case of removable containers, since no restriction was placed on the order of the collection points, a rotation schedule could be established by a simple combination of operation times.

The use of removable containers shortened the time required for tasks at collection points. However, the introduction of the existing type of removable container that allowed collection of a larger number of discarded tires did not directly improve the efficiency of collecting discarded tires.

As the measure of the small number of discarded tires, the development of a new type of removable container as in Fig. 7 that can be rapidly delivered to the next collection point through the simulation is considered in the scenarios.

5 Analysis of the Delivery System of Fuel Chips

5.1 Outline of the Simulation Model

As Fig. 8 shows, the simulation model used in this study is composed of three echelons: the intermediate treatment factory, the physical distribution complex, and the destination. The management situation of company A's delivery system for fuel chips and the conditions set for the scenario analysis allows use of a simulation model to investigate the following three desirable situations, from the viewpoints of the environment, load, and cost of the entire delivery system:

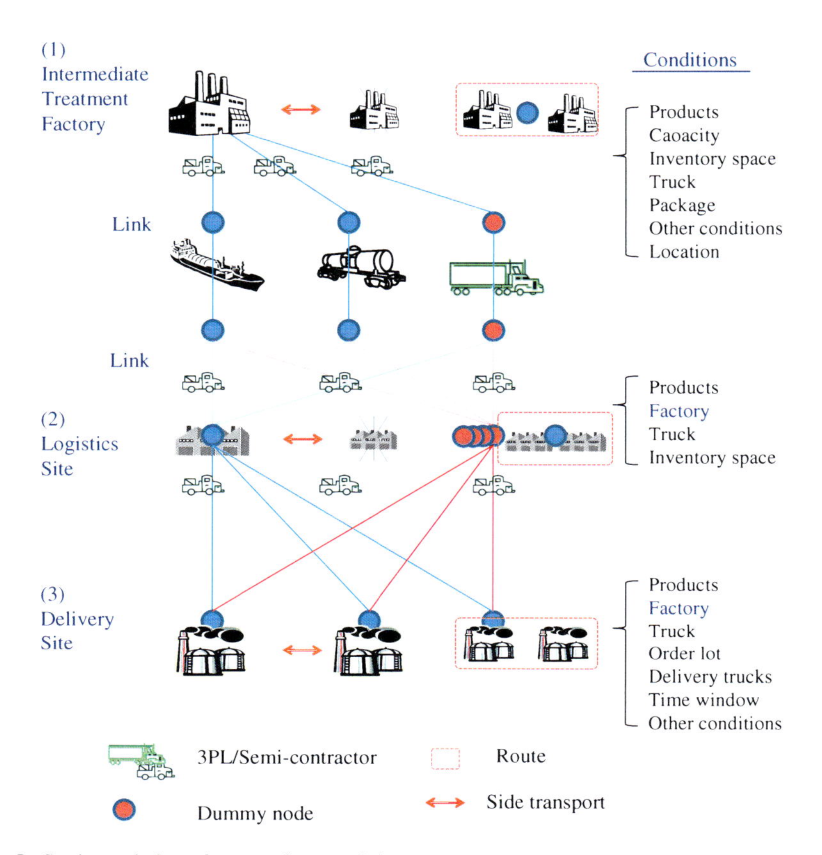

Fig. 8 Setting echelons for recycle materials

- Desirable situations of complexes (nodes) belonging to each echelon, i.e., location for Intermediate treatment factory (Fig. 8a), location for Logistics site (Fig. 8b), and location for Delivery site (Fig. 8c)
- Desirable situation of movement (links) between complexes belonging to the same echelon, i.e., side transport between Logistics site and Delivery site ((b) and (c) in Fig. 8)
- Desirable situation of movement (links) between complexes belonging to different echelons, i.e., transport mode between Intermediate treatment factory and Logistics site

By using this simulation model, which expresses the node locations in (a) Intermediate treatment factory, (b) Logistics site, and (c) Delivery site, and links between the nodes over echelons, the expended costs are calculated.

The actual investigation is performed as follows: The intermediate treatment is performed on collected discarded tires, the treated tires are recycled into fuel chips, and these are transported and distributed to cement factories, paper and steel mills as valuable materials. For transportation and distribution to the mills, the effects of adoption of a modal shift are examined with respect to reduction in environmental load.

The modal shift (Lowe 2005) combines truck, marine, and rail transportation. The location problem, including elimination and consolidation of intermediate treatment factories and facilities possessed by the factories, are also reexamined while considering both the collection system for discarded tires and the delivery system for the fuel chips. In this study, only the combination with trucks and ships is examined.

5.2 Numerical Results for the Delivery System

The delivery of produced fuel chips currently considers the following three scenarios for its improvement:

- Transportation and distribution performed only by truck
- A modal shift, combined with truck and ship transport
- A hybrid system: a combination system with truck direct deliveries and the modal shift system

For example, concerning the Tohoku district in the first scenario, 1,000 ton of fuel chips of discarded tires are delivered monthly by a 20-ton truck; in the second scenario, 1,000 ton of fuel chips of discarded tires are transported by sea once a month, and delivered to 27 cement factories and paper and steel mills, using a milk run delivery system. The third scenario is the hybrid combination system which depends on distance and delivery time.

The delivery area of Company A, located near Tokyo in the Kanto district (metropolitan area), can be seen in east Japan, 500–700 km from their intermediate treatment factory.

Currently, Company A is considering three scenarios for cost-reduction measures: a truck direct delivery, a modal shift system, and a hybrid system. As for the first scenario (only by truck), the fuel chips are delivered in small amounts, 5–10 tons, by truck. Figure 9 shows the image of the truck direct delivery.

However, by reducing the number of trucks (and therefore transport costs), an effective improvement measure is suggested. That is the second scenario: introducing the modal shift by the delivery of a larger amount, 1000 tons by sea, utilizing a sea port depot from which the area deliveries are made by 20 ton truck (See Fig. 10).

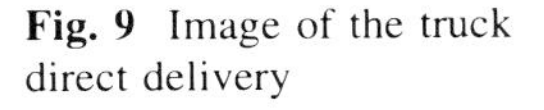

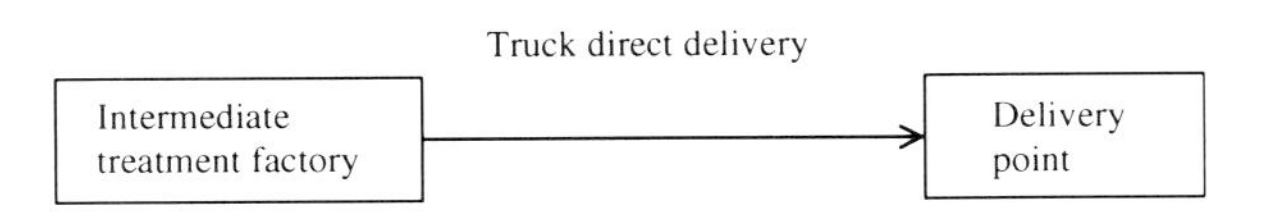

Fig. 9 Image of the truck direct delivery

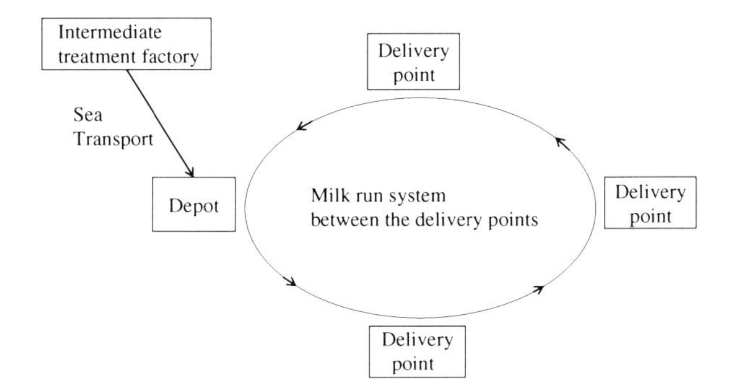

Fig. 10 Image of the modal shift system

That is, the truck-only delivery should be changed to include sea transport from the Kawasaki Sea port, the metropolitan sea port delivers to the optimized sea port in northeast Japan. The best candidate is the sea port in Shiogama, Sendai, chosen from the interviews with people involved in the fuel chip industries.

The modal shift system introduces the use of the depot and the milk run system among which the delivery points can be found.

The third scenario is a hybrid system, which is a combination system of truck direct deliveries and the modal shift system, which is determined by distance and delivery time. Figure 11 shows the image of the hybrid system.

Considering the scenarios, the case study of the deliveries from Company A in northeast Japan can be considered.

Company A's current delivery system parameters are shown in Table 6. The company has 27 delivery points in northeast Japan. The fuel chips are transported by sea from the intermediate factory to the depot where they are provisionally stored. Then, delivers of 5–55 tons are made once or twice a week to each delivery point.

Figure 12 shows the delivery area of company A, including its delivery points, intermediate treatment factory, and sea port

Figure 13 shows the scheme of the simulation of the truck-only delivery from the metropolitan area to the northeast of Japan. The "X" marks in Fig. 13

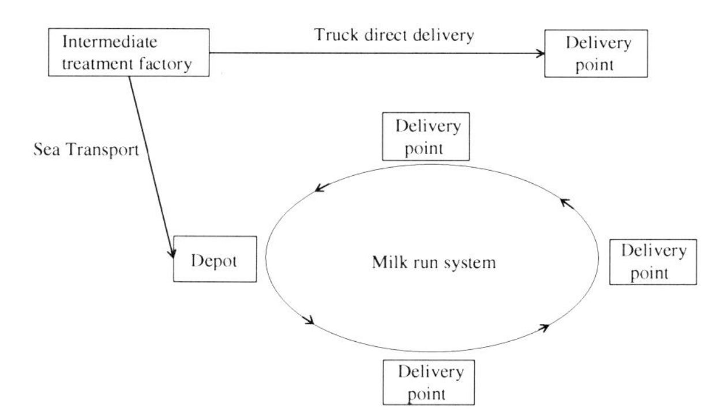

Fig. 11 Image of the hybrid system

Table 6 Condition of the simulation

Items	Contents	Memo
Delivery points	27 points	In the northeast Japan
Supply lots for a point	5–55 tons/week	————————
Frequency for a point	1–2 times/week	————————
Truck loading lots	20 tons/truck	No limitation for the transport
Ship loading lots	1000 tons/ship	No limitation for the transport
Intermediate treatment factory	Located in Kanagawa Prefecture, Japan (the metropolitan area)	————————-
Candidate depot	Near the sea port in the northeast area of Japan	Chosen by the interviews from the people concerned

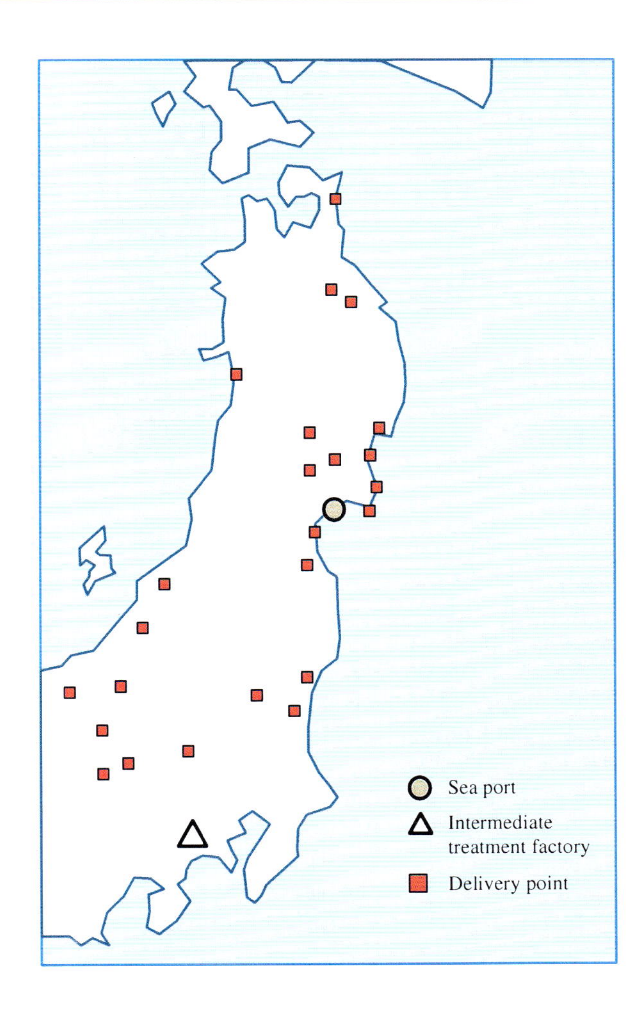

Fig. 12 Delivery area

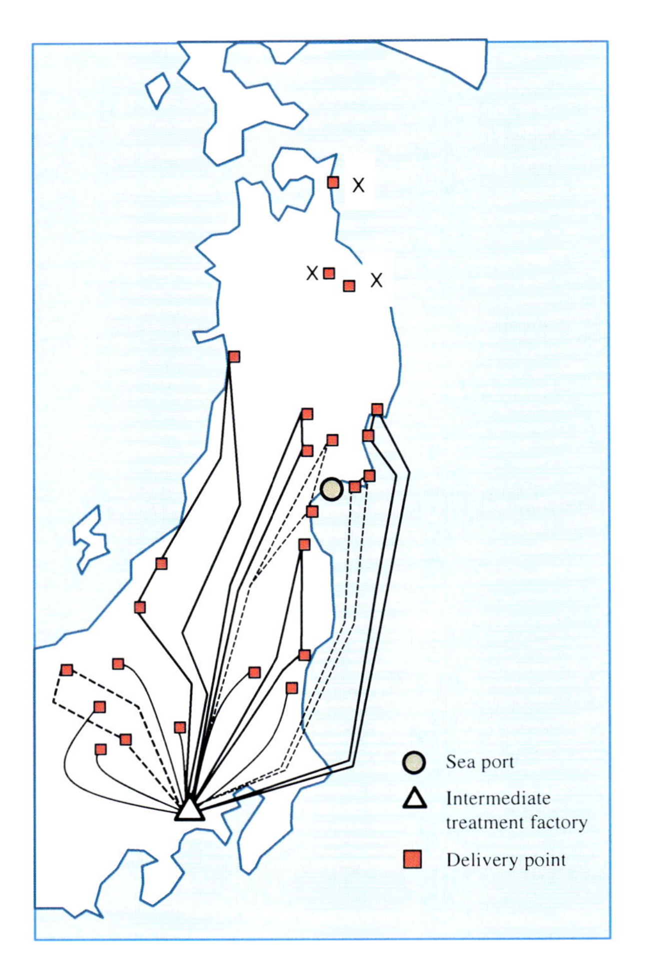

Fig. 13 Truck direct delivery area

designate areas that are impossible to transport to within the legal daily working time for the truck drivers, that is, the deliveries require more than one day for driving, which is not effective for logistic systems.

Consequently, it is suggested that the depot be set at a sea port in east Japan in order to receive large shipments of fuel chips from the intermediate treatment factory and a milk run delivery system is introduced from this depot to the delivery points.

Clearly, there is a limit for truck-direct deliveries from the intermediate treatment factory.

Improving the efficiency of the truck direct delivery system by introducing the modal shift system is one possible solution.

Figure 14 shows the scheme of the introduction of the modal shift system from the metropolitan area to the optimized sea port in northeast Japan. All the fuel

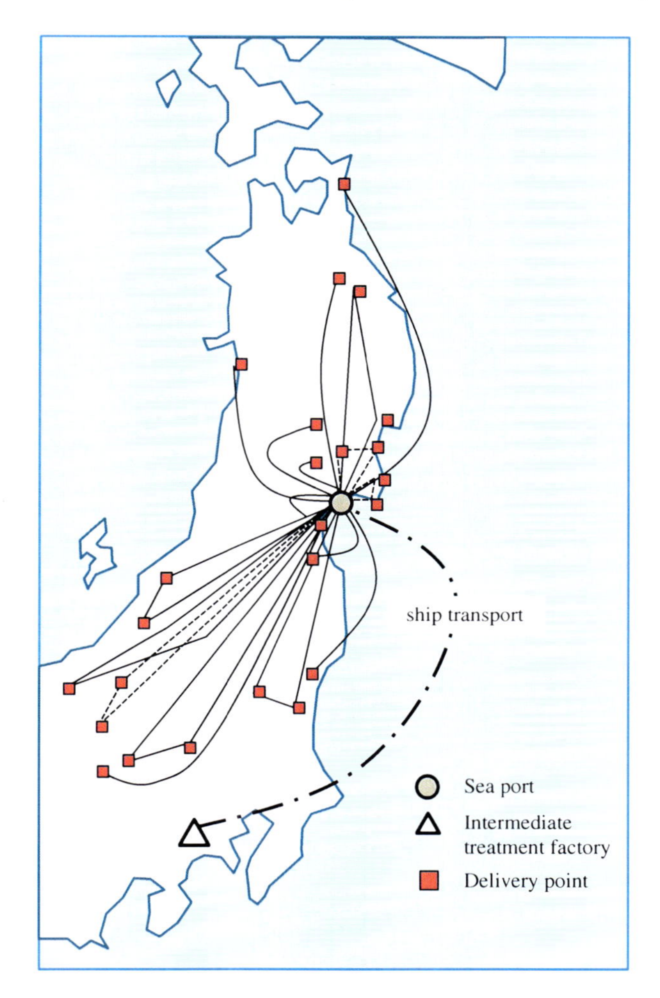

Fig. 14 Introduction of the modal shift system

chips are transported to the delivery points by sea and are then delivered through the depot with a milk run system.

However, another option can be suggested. That is the third scenario, as shown in Fig. 15. For the closer delivery areas (the area northwest from the intermediate treatment factory), a truck direct delivery may be effective, while the northeastern area (north of Fukushima where the truck driving time is too long for the drivers) can be delivered via optimized sea port and by truck. That is the hybrid system: a combined system of truck direct delivery and the modal shift system.

Table 7 shows the results of the simulation compared with the modal shift and the hybrid system. The results of the truck direct delivery system cannot be shown because of the existence of impossible delivery points.

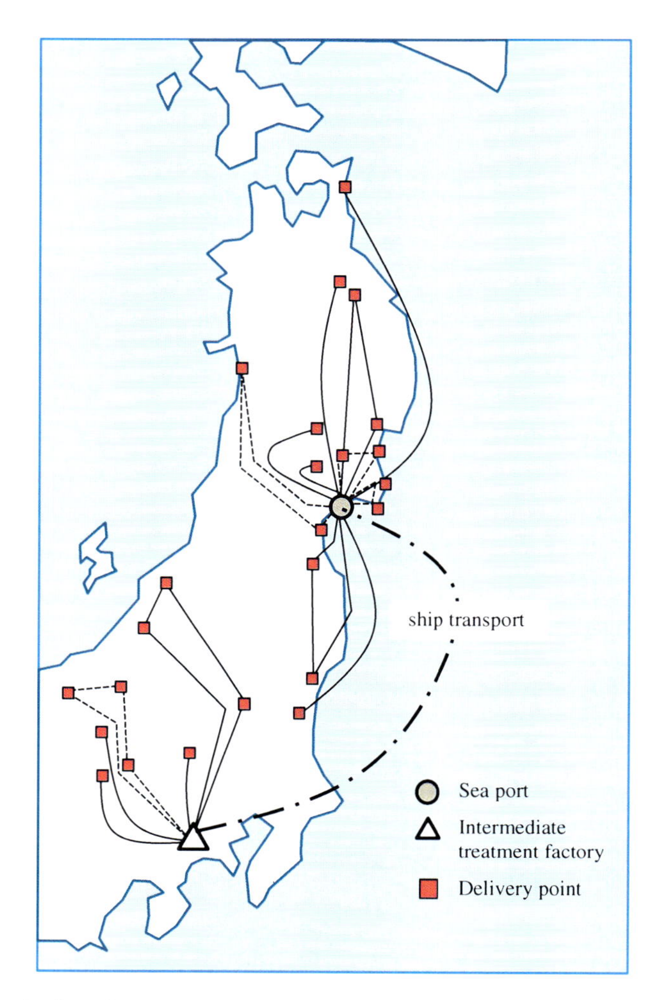

Fig. 15 Introduction of the hybrid system

Table 7 Cost Comparison

Measure	Cost (by truck)	Cost (by sea)	Cost (for depot)	Total
Modal shift	3,841,380	6,200,496	864,000	10,905,876
Hybrid system	2,994,940	3,100,248	432,000	6,527,188

It is clear, however, that Company A can reduce its total transport cost by more than 60 % through the introduction of a modal shift with a depot and milk run system, compared to the truck-only delivery system.

As for the comparison with the modal shift and hybrid systems, a great difference between the two measures can be seen. By the hybrid system, the cost (by truck) can be reduced because of depot efficiencies. In addition, both the truck and sea transport costs can be reduced by introducing the hybrid system, which especially can improve the efficiency of sea transport.

Consequently, by introducing not only the truck direct delivery system, but also the modal shift system, the impossible delivery points (those that are too far for one-day driving deliveries) are avoided. Setting the depot in the northeast area should be effective for this solution.

6 Conclusion

This study presents a total reverse logistics solution for the collection and transport of discarded tires from retail stores to the delivery of fuel chips made from the discarded tires.

As for the collection and transport, it is important not only to collect tires from a wider area, to introduce a cooperative collection system and extension of the collection area for discarded tires, and to increase in the number of collection points for an integrated intermediate treatment factory, but it is also necessary to introduce other cost-reduction measures: introduce an RFID system and an ASN system, develop an electronic manifesto, and utilize removable containers.

Concerning the improvement of the delivery system, compared to the current process of delivering only by truck, a new hybrid system should be introduced. The new system would be a combination of a modal shift system and the traditional truck direct delivery system.

A cooperative reverse logistics network will require the establishment of a new scheme that utilizes accumulated knowledge in forward logistics.

The improvement measures for an effective reverse logistics system for discarded tires have not been clearly shown in previous studies.

In this study, however, the collection and transport network and the delivery system of thermal products recycled from tires are systematically covered, viewing the total solution. Finally this study suggests a direction for the effective improvement for the reverse logistics system for discarded tires as shown in Fig. 16.

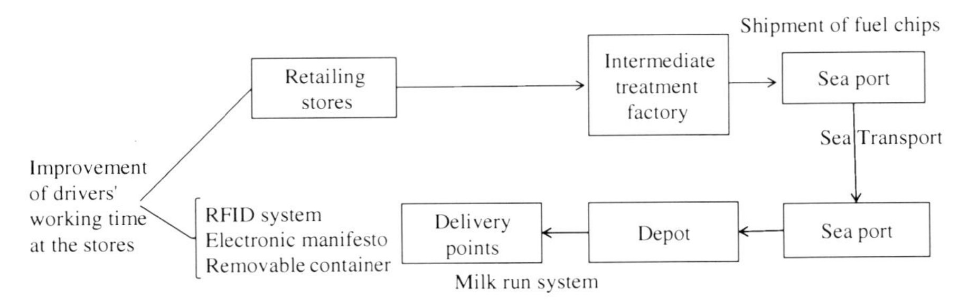

Fig. 16 Total scheme of the effective reverse logistics system for discarded tires

In the future, the delivery system for fuel chips should also be analyzed to determine the possible reduction in environmental load, and should include a number of factors including collection complexes in overseas regions such as China and South Korea in addition to Japan, and consumption areas for the recycled goods.

Also, concerning the improvement of the delivery system, from the actual scheme, just only delivery by truck, the scheme should be changed into the hybrid system, which is the combination with a modal shift system and truck directly delivery system.

A cooperative reverse logistics network will require the establishment of a new scheme while utilizing accumulated knowledge in forward logistics.

In the future, the delivery system for fuel chips should be also analyzed for a reduction in environmental load, and should include a number of factors including collection complexes in overseas regions such as China and South Korea in addition to Japan, and consumption areas for the recycled goods.

References

Beullens P, Van Oudheusden D, Van Wassenhove LN (2004) Collection and vehicle routing issues in reverse logistics, reverse logistics. Springer, New York, pp 95–134

Fleischmann MJ, Bloomhof-Ruwaard M (2004) Patrick beullens, and Rommert Dekker, reverse logistics network design. Reverse Logistics, pp 68–69

Kubo M, Kogaku L, Shoten A (2001) pp 111–116

Lowe D (2005) Intermodal freight transport. Elsevier, Oxford, pp. 6–10

Smichi-Levi David, Kaminsky Philip, Simchi-Levi Edith (2002) Designing and managing the supply chain. Japanese Translation, Asakura Shoten, pp 70–71

Suzuki K (2010) Green supply chain no Sekkei to Kochiku. Hakuto Shobo, Tokyo p 13

Toth Paolo, Tramontani Andrea (2008) An integer linear programming local search for capacitated vehicle routing problems. The vehicle routing problem. Springer, New york, pp 279–280

Part III
Recovery of End-of-Life Vehicles

The Necessity of Recycling Networks for the Sustainable Usage of Automotive Parts: Case Study Germany and PR China

Alexandra Pehlken, Wolfgang Kaerger, Ming Chen and Dieter H. Mueller

Abstract The aim of this chapter is to point out, that developed processing of end-of-Live-Vehicles has a high potential for sustainability. We present existing networks for used automotive parts with the case study of the callparts network in Germany. Callparts Recyling GmbH provides a network for the automotive industry to guarantee a high standard of end-of-life vehicle recycling. Their main interest lies on the distribution of used automotive parts. The potential for sustainable resource management will be shown accordingly to the Chinese ELV market. Used automotive parts are kept in a closed loop recycling system and are further used as automotive parts by showing a good overall lifecycle performance. This implies savings of primary resources on material and process level. Among recycling hierarchy the reuse of used parts always achieves the most favorable eco-balance and thus, gains the highest priority. The legal situation in China has to pave the way for tapping their potential for saving materials and energy in the future through car parts remanufacturing. By setting up a nationwide recycling network we might achieve the goal.

Keywords Remanufacturing · Closed loop supply chain · Recycling network · Database

A. Pehlken (✉) · D. H. Mueller
Bremen University, Badgasteiner str1 28359 Bremen, Germany
e-mail: pehlken@uni-bremen.de

W. Kaerger
Callparts Recycling GmbH, Gewerbegebiet 14669 Ketzin, Germany

M. Chen
School of Mechanical Engineering, Shanghai Jiao Tong University, Dongchuan Rd. #800 200240 Shanghai, People's Republic of China

P. Golinska (ed.), *Environmental Issues in Automotive Industry*, EcoProduction, DOI: 10.1007/978-3-642-23837-6_9,

1 Introduction

The EU Disposal of end-of-life Vehicle (ELV) Act specifies quotas of material recycling (EU 2000). The ELV Directive 2000/53/EC established goals to minimize the effect of ELV's by setting recycling, reuse, and recovery targets for the materials used in all manufactured vehicles. The directive requires that 95 % of ELV waste must be reused by 2015, with only 10 % of this recovered through energy. In addition to pure recycling of materials, the reuse of components and subassemblies makes sense not only in line with the Lifecycle Management and Waste Act (KrW-/AbfG 1994) from Germany but also from the economic perspective.

The reuse of automotive parts and subassemblies necessitates not only an unambiguous identification and classification of the extracted parts, but also a comprehensible documentation and description of their present condition, as automotive parts and subassemblies have been produced by different manufacturers and have been extracted from different vehicles. Consequently a reliable classification and identification is necessary to gain the acceptance of the customers. The present VDI-guidelines VDI 4080, 4081, and 4082 "Recycling of cars" are limited to a description of the treatment process of used automotive parts [VDI 4080–4082]. However, representative quality methods to anticipate the future life-time of recycled automotive parts and guarantees are still missing. Due to the increasing market for cars and also the increasing amount of vehicle types, not only for cars but also of different components per type, it is necessary to establish an excellent data network between different recycling companies on the one hand and, on the other hand, to establish competence centres for quality analysis to improve the contingent and market for reused parts.

A system like recycling stock exchange is the basis for a well operated network and raises the chances for being successful. In Germany exists a well operated Callparts IT-network which has already shown an available option for automotive components reuse. The distribution of automotive parts is sustained by providing a website and developed software. The software provides information on the logistics as incoming vehicles, market place for residues, storage and sale of automotive parts. Through the software all partners are linked and their virtual storage is made visible on the website.

Very important and the basis for a market of reused/remanufactured components represents classification regulations to specify the quality like vehicle bodies and engines/transmissions for example. Quality methods help to raise the acceptance for used automotive parts and increase the reliability of these parts (Pehlken and Müller 2009).

2 Automotive Car Parts in Germany

In Germany § 4 of the Lifecycle Management and Waste Act (KrW-/AbfG 1994) clearly indicates that all kind of waste materials has to be avoided before it comes to recycling procedures. Remanufactured parts are reused for their original purpose and perform the best life cycle compare to most recycling processes. The "Altfahrzeug Verordnung" (Altfahrzeug 2002) regulates that all car manufacturer in Germany are obliged to take back their scrap cars and recycle them in dismantling facilities. 95 % of a car has to be reused AND recycled by the year 2015.

The German car market has a volume of around 45 million cars. The average life cycle of cars is around 8 years. Especially those which are older than 3 years are suitable for used spare parts demand.

Understanding the flow of ELV (see Fig. 1), market players and competitors are examples for basics to be in favor of managing a network. The ELV-Flow under economical and legal factors affects the input of a network.

Since Germany is very ambitious on technical level and quality of cars. This also leads to a special market situation. For local use are high quality used spare parts necessary for the German market. However Germany represents for cars and also for spare parts an export market.

Germany exports high quality used spare parts to those neighbors who buy the younger used cars and the necessary spare parts from the German market. For example European neighbors, Eastern Europe and the Middle East.

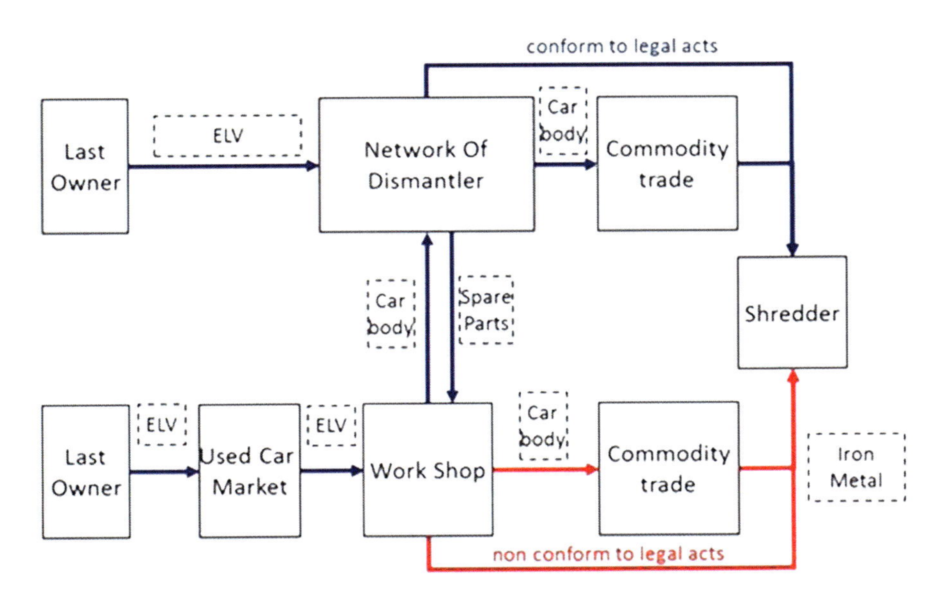

Fig. 1 Flow of ELV under controlled or less developed conditions

Low quality and lower price level is expected to those countries which import older cars and older used spare parts like African countries, Arabian countries but also Bulgaria and Romania.

The sales of used spare parts is performed by car dismantlers and used spare parts dealers who act against retail markets with commercial clients (repair shops, garages) and private clients (who buy only 1–3 times in their life).

The sales media are

- the counter of dealers and dismantlers,
- the prolonged counter of dismantler networks with virtual storage,
- and the internet platform which are maintained by car dismantlers.

To improve the used spare parts sales the policy has to

- present used spare parts in a single form (not as bulk ware),
- present spare parts on high quality level with a qualified data base and pictures, and
- use dismantler networks or virtual stocks to meet the clients demand as much as possible.

Reused parts have to follow quality guidelines as well. The VDI Guidelines "Recycling of cars" VDI 4080, 4081 and 4082 specifies minimum criteria for the qualitative description of used car parts (passenger cars, commercial vehicles and motorcycles). The highest quality level represents class A, followed by class B and C. The quality of class A is comparable to new automotive parts with nearly no minor functions, whereas class C represents the car parts with functional defects. Table 1 shows descriptions of class A and class C used engines according to the VDI guideline. Proper handling of used parts includes, moreover, clear labeling, well storage and packaging. Used car parts come from ELV without further reconditioning or remanufacturing. In accordance remanufactured parts are used parts that are remanufactured according to the state of the art. They must be comparable to the corresponding of a new part in terms of quality and function.

Table 1 Example for quality classes for used engines [VDI, 4080]

Component	Quality class A	Quality class C
Engines	Mileage passenger car <75.000 km and age to 5 years	Mileage passenger car from 150.000 km
	Commercial vehicle <200.000 km	Commercial vehicle from 700.000 km
	Motor cycle <l20.000 km	Motor cycle from 40.000 km
	Functional or performance test; no recognizable defects	Functionality
	Good compressions	
	No unusual sounds	Unusual noises
	No leaks at the gaskets	Leaks
	History known	History unknown or incomplete

Table 2 Data records for car parts as EDP interface [VDI 4081]

Meaning	Type of format	Brief description
No. of car manufacturer	Numerical	Identification of manufacturer
No. of car model	Numerical	Identification of model
No. of car types	Numerical	Identification of vehicle
VDI part No.	Numerical	Identification of part
Supplement or part No.	32 charchters	Field no currently in use; intended for supplementary identification of part
Location of installation	32 characters	Identification of location of installation by combining criteria
Quality	Numerical	Specification of quality class (see VDI 4080)
Date of first registration	7 characters	Month two digit number/Year three digit number
Motor code	32 characters	Motor code
Transmission code	32 characters	Transmission code
Quantity	Numerical	Quantity
price	Numerical	Supplier's price w/o additional charges (shipping, commission, insurance)
Shipping cost	Numerical	Shipping cost
currency	3 characters	Default setting: Euro
Type of insurance	Numerical	0 = uninsured; 1 = insured by recipient of data; 2 = insured by supplier
Comment	255 characters	Available for individual comments
Upload ID	16 characters	Unambiguous label allowing to idetify source of data and upload. This allows unambiguous allocation of later uploads

To be "comparable" a standard has to be set. The offer and supply of reused and remanufactured parts and subassemblies removed from end-of-life vehicles has to be developed with the requirements in Table 1.

The list of recycled parts compiles all parts disassembled during the recycling process. This list and the coding of the parts serve for unambiguous identification, creating the basis for data exchange using electronic data processing (EDP) interface for recycled car parts data records. The disassembling company has the responsibility to make a decision if disassembly and exploitation of these parts are reasonable. Table 2 shows data records for recycled car parts, which using several parameters by means of the EDP interfaces.

According to the available information necessary data on the components and their registration document, number of previous owners, environment of use, type of use, maintenance and repair documentation are listed. Every car part is labeled with the necessary information and provided within the network (Fig. 2).

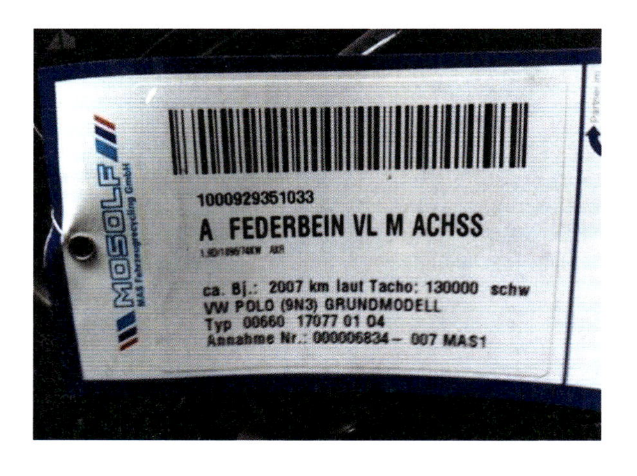

Fig. 2 Example of component identification code

3 Network in Germany for Used Car Parts

The Callparts Systems is one of the important players in the field of used spare part marketing on high quality level in Germany.

Callparts was founded in 1993 in accordance with the European Institution and the implementation of the waste management systems in the EU and their member states. They were focused on ELV-treatment and-organization.

Callpart's idea was to implement 80 partner networks in Germany to take back and treat ELVs in an adequate manner, which is achieved in the meantime. Priority has had the reuse of spare parts out of dismantled cars. Second priority according to the waste pyramid was to reuse secondary raw materials and save resources.

Nowadays Callparts Recyling GmbH provides a network for the automotive industry to guarantee a high standard of end-of-life vehicle recycling. Their main interest lies on the distribution of used automotive parts. The CallCar network was implemented in 2004 with the participation of two automotive producers, automotive import companies and about 100 vehicle dismantling companies (see Fig. 3). Additionally, they possess a dismantling plant near Berlin with a capacity of dismantling 15.000 cars per year. Main focus lies on the extraction of high valuable materials or components from end-of-life vehicles for further use (Kaerger 2010). No other company is providing their own dismantling company and their own network within the automotive industry. The advantage is that all information about automotive parts is directly available without any filtering by other providers.

Callparts System provides a network for dismantling companies (Fig. 4). The distribution of automotive parts is sustained by providing a website and developed software. The software provides information on the logistics as incoming vehicles, market place for residues, storage and sale of automotive parts. Through the software all partners are linked and their virtual storage is made visible on the website www.callparts.de (Fig. 5).

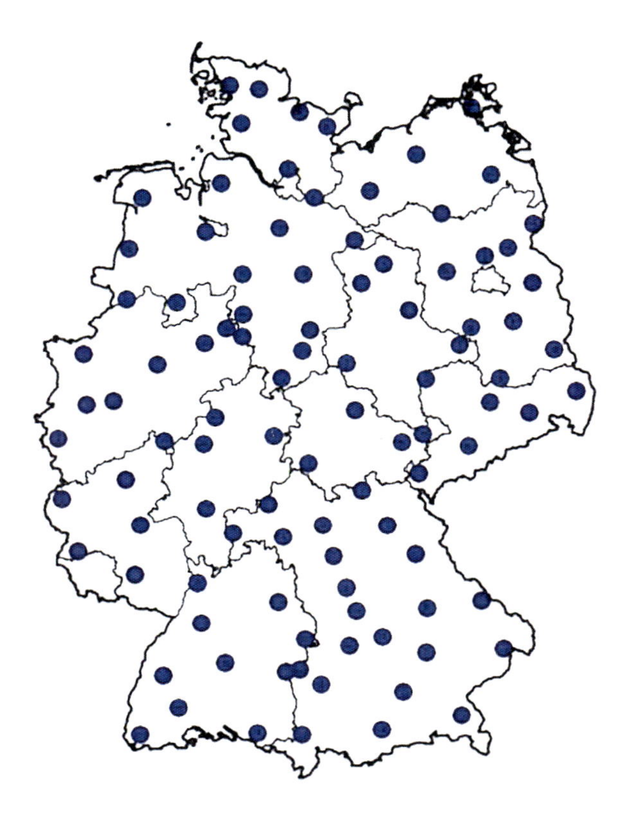

Fig. 3 Callcar recycling network

The network system in Germany can be extended with quality assets like description and function of automotive parts. The components which are good for re-use or remanufacture respectively are being identified in the country of origin. After being transferred into a database, the information is available all over the country in the world-wide-web.

Quality statements have to be developed and when possible, physical descriptions have to be identified and quantified, like (mainly based on VDI 4080) vehicle bodies, engines, other units, electric parts, interior equipment, and identification of valuable recycling components (like catalysts).

Unambiguous quality statements are necessary to improve the acceptance of remanufactured components; this involves an evaluation on the automotive parts potential for being remanufactured. The intended network will increase the recycling potential of automotive parts because automotive parts (1) will be listed (2) will be made available for an international market and (3) besides remanufacturing, also further possibilities of re-use and material recycling are considered.

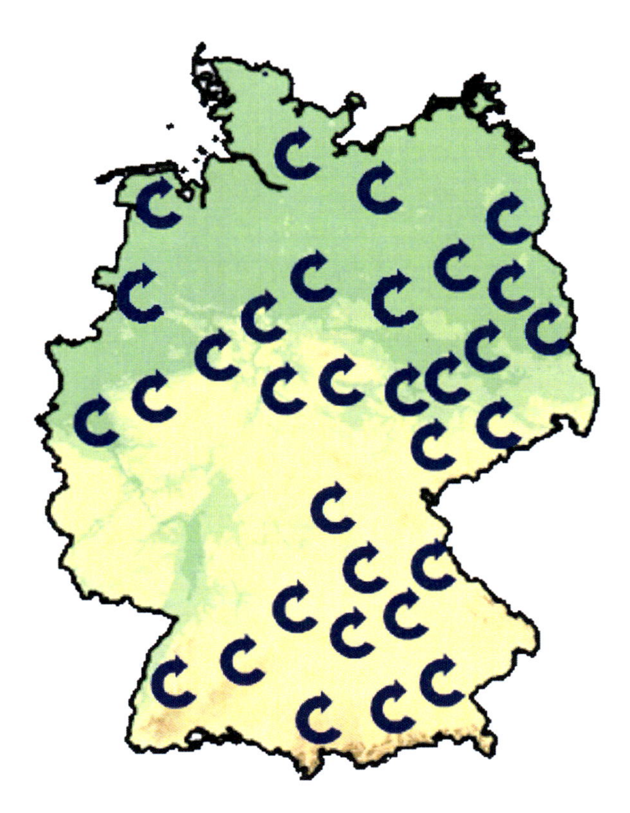

Fig. 4 Callparts network for ELV

4 Situation for Used Car Parts in PR China

4.1 History of Car Sales and the Need for Used Car Parts

In recent years, vehicle population has increased annually in China. According to statistical data from the China Association of Automobile Manufacturers, the vehicle market in China set a record with sales of approximately 13.791.000 units in 2009, bringing the total number of vehicles to 62.800.000. In 2010, China became the global leader in automotive production and consumption after surpassing the United States. The total number of automobiles reached 75 million at the end of 2010. According to conservative estimates, the total number of automobiles in China is expected to be over 490 million in the future (Wang and Ming 2011). The Table 3 shows the forecast of vehicles in China; the statistical data was provided by the China Automotive Technology and Research Center (CATARC). In general, the volume of ELV increases with the rise in vehicle sales. The rapid growth of vehicle population in China would cause subsequent environmental problems if not addressed in advance.

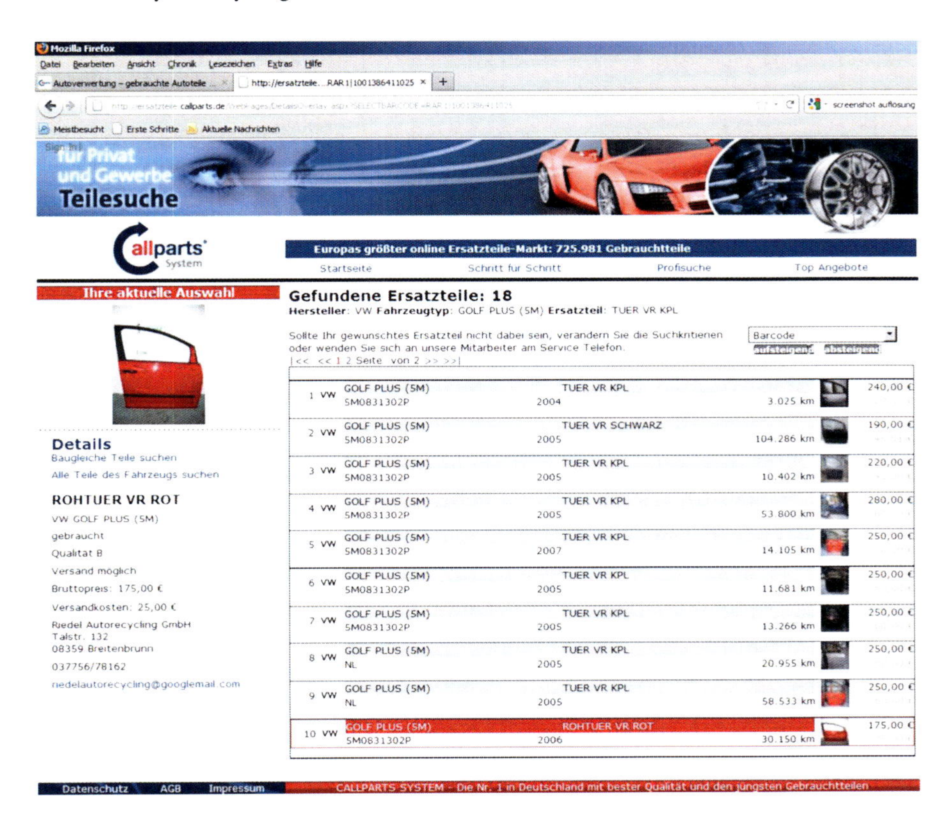

Fig. 5 Online-view into parts stock

Table 3 Forecast of vehicles in China (Wang and Ming 2011)

Years	Vehicle population (million)	Vehicle sales (million)	Total number of scrapped vehicles (million)	Ratio ELV/vehicle sales (%)
2015	95,38	14,91	6,44	6,75
2017	112,72	16,91	7,78	6,90
2020	141,03	20,05	9,95	7,06

Counting solely the amount of secondary resources that are going to be available through ELV recycling in China 10–15 % of the national scrap steel is provided each year. With the rising amount of ELV vehicles the available scrap steel is estimated for 15–22 million tons in 2020. The number for non-iron metal scrap can be predicted to 2.2 million tons in 2020 (Rueth 2011). The numbers for used car parts that are still functional and able to be reused cannot be estimated today but the value of these parts are much higher than the number for scrap metals. Additionally, the life cycle performance of these reused parts is always rated higher than common recycling processes.

4.2 Current Situation and Necessary Research

Due to the extremely increase of new car sales in China the interest in reused automotive parts is growing. As different automotive parts do not get unusable at the same time, recyclable, respective reusable, automotive parts achieve high popularity in China. Consequently there is a strong demand for automotive parts in use. Already more than 450 qualified dismantling companies have emerged their existence on this market. It is not a question whether automotive parts derive great value as resources, because they also hold residuals like catalysts, which need to be recycled mandatorily. The implementation of the EU end-of life vehicle (ELV) directive had a profound influence on China's automotive industry, leading to the consideration of concepts -such as extended producer responsibility (Zhang et al. 2011) The establishing legislation and specification system on automotive products recovery could be regarding as response of China towards EU ELV directives, from national law to detailed technical specifications, and from vehicle design to ELV recycling (Xiang and Ming 2010). The actual laws, policies and guidelines are as follows:

(1) National Law

- "Law of the People's Republic of China on Prevention of Environmental Pollution Caused by Solid Waste" The Law is enacted for the purpose of preventing and controlling environmental pollution by solid waste, safeguarding human health, safeguarding the ecological environment and promoting the sustainable development of economy and society.
- "Law of the People's Republic of China on Circular Economy Promotion" The law, designed to provide a legal framework for its national sustainable development strategy, is goal at bringing attention to a wide range of environmental considerations as China continues its high-speed economic growth. The law requires low energy consumption and high resource efficiency, low emissions of pollutants, and minimal waste discharge, using the principles of reduce, reuse and recycle. The law also requires the recycling of larger waste products such as end-of life vehicles and ships, mechanical and electrical products. Incentives and penalties are included.

(2) National Technical Policies

- "Automobile Industry Development Policy" released in May 2004
- "Automobile Trade Policy" released in August 2005
- "Technical Policy of Automotive Products Recovery" released in Feb. 2006
- Administrative Measures for Pilot Remanufacturing of Automobile Parts and Assemblies" released in March 2008
- "Management Regulation of ELV Take-back", a revised Decree 307, draft for review
- "Administrative Rules of Recoverability Rate and Forbidden Substances for Motor Vehicles", draft for review

(3) Technical Specifications

- GB/T 19515-2004 "Road vehicles-Recyclability and recoverability-Calculation method"

- HJ 348-2007 "Environmental Protection Technical Specifications for Disassembly of End-of-Life Vehicles"
- QC/T 797-2008 "Material Identification and Marking of Automotive Plastic, Rubber and Thermoplastic Elastomer Parts"
- GB 22128-2008 "Technical specifications for End-of-Life Vehicles recycling and dismantling enterprise"
- Twelve specifications for automotive parts and assemblies remanufacturing

The legal system of ELV recycling in China comprises three levels: national laws, national technology policies and technical specifications. At the year of 2001, Decree #307 "Management Measures on Take-Back of Scrapped Automobiles" about ELV disposal was issued by State Department of P. R. China. This is the highest law concerning about the disposal of ELV in China today. In 2006 the "The Motor Vehicle Product Recovery Technology Policy" was issued by the National Development and the Reform Commission (NDRC), the Ministry of Science and Technology and the State Environmental Protection Administration (SEPA). According to the regulation, Chinese automobile manufacturers and importers will be responsible for collecting and recycling vehicles from 2010 or designating other authorized end-of life vehicle dismantlers to handle the business. The Chinese policy requires that vehicle manufacturers recycle at every phase of the products life cycle (e.g. manufacture, sales, repair, maintenance, dismantling, etc.) to help attain a recoverability rate of 95 % on automotive products which will be manufactured and sold in China by 2017 (Ming 2009). The new regulation was designed to encourage domestic automobile producers to use more environmental benign materials. The new policy will surely raise the costs to automobile manufacturers and importers and poses a higher demand on technology innovation. But over the long term, it will help to increase the sustainability of the Chinese automobile industry.

According to Chen Ming there are about 492 qualified ELV dismantlers approved by the state administrative department, and more than 1772 take–back stations in Chinese cities. This results in a capacity of 1.2 million vehicles per year. Dismantling in China must be consistent with China's national conditions since China has abundant labour resources. Therefore, the dismantling operation mode in China is a combination of mechanization tools and manual dismantling.

A convenient infrastructure is requested, because in every province in China workshops have to deliver both: absorbing all re-usable cars and providing the automotive parts for the market. Those parts can be served on demand by the convenient delivery-services. Until now China has only few dismantling companies. However, they act decentralized. At this time, the increasing number of car sales show a lack in the support of re-usable automotive parts, as demand is even higher. Figure 6 shows the concept of a pilot recycling system for used car parts in Shanghai to cope with the rising demand for uses car parts.

We observe different economic development of eastern, central and western regions of China. Prospective it will be the characteristic Chinese economic development. The regional difference in economic requirements and consumptions leads to the obvious difference in vehicle consumptions. The new vehicles begin

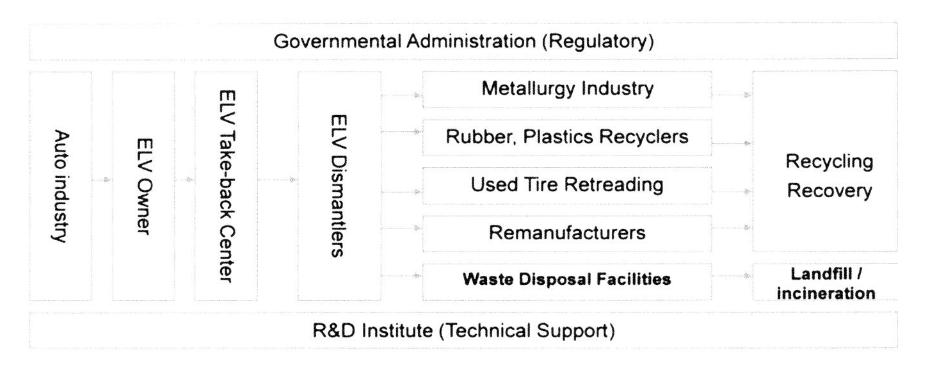

Fig. 6 A pilot recycling system for used auto parts in Shanghai, China

their life cycle in eastern and their end of life in western. The development and layout of Chinese vehicle recycling industry should consider this social fact. A model of vehicles recycling and reverse logistic infrastructure in China will consider this fact:

(1) In central and western regions of China, a lot of ELV dismantlers should be established and accept obsolete vehicles. After being dismantled, the auto bodies, tires, plastics and used oils should be treated locally. For instance, bodies are recycled as scrap steel; tires and plastics are recycled as crushed crumb. The effective and low-cost process and equipment should be developed and applied in dismantlers of these regions.
(2) The repairable parts and assemblies which are collected in dismantlers or service stations should be sent to remanufacture industry after gathering of a certain amount. The remanufacturing enterprises could be specialized or comprehensive, and could be independent or belonged to vehicle manufacturers. The remanufacturing parts came into spare parts market. The failure parts in service stations are replaced and treated as the same as those in dismantlers.
(3) Dismantlers collect reusable parts directly and send them to original equipment manufacturers (OEM) or vehicle manufacturers respectively via logistics between service stations and vehicle manufacturers. The automotive suppliers should establish their specialized workshops or departments for dealing with those parts which could be reassembled after strictly quality checking and testing (Ming 2005).

5 Conclusions and Further Research

Our aim is to receive valuable resources from End-of-Life vehicles (ELV). After their dismantling, the automotive parts get classified and catalogued. Among recycling hierarchy the reuse of used parts always achieves the most favourable eco-balance and thus, gains the highest priority. On the same level we identify the remanufacturing of automotive parts. In order to give rise to a structured recycling process, it is intended to further develop the network based upon information on dismantled automotive parts. For this purpose, it has to be distinguished between the input (complete cars) and the output (single automotive parts) as a basis for the network.

Main objectives can be identified as:

1. Set standards and procedures for improving quality measurements like vehicle bodies, engines, transmissions, steering gear, electrical parts and others
2. Raise the market acceptance for remanufactured or reused car parts through networking in redemption and marketing communication (take-back network is highly developed in order to get as much end-of-life vehicles as possible)
3. Increase efficiency of logistics, transport and marketing (network).

The basis for a market of reused/remanufactured components can be represented by classification regulations to specify the quality like vehicle bodies and engines/transmissions for example.

The potential for sustainable resource management is very high. Manufactured products are kept in a closed-loop recycling process. Therefore, a car part is used as the same car part. Remanufactured products have a good life cycle performance as well because they are able to reach the quality standards of new manufactured products. This implies that resources (on material and process level) are kept to a minimum (Pehlken et al. 2010). The implemented network on car parts distribution guarantees an ideal communication platform between customer and product. Therefore, the access to all available parts is easier and makes the further reuse possible. The development of a new quality method for used combustion engines implies a reliable automotive parts reuse.

The network is already running successful for many years in Germany and has been implemented in the Czech Republic in 2005. China' vehicle remanufacturing industry is still in its early stage. In terms of policy China must strengthen further its implementing rules, regulation, and standards related to vehicle recycling and remanufacturing.

References

Altfahrzeug V (2002) Altfahrzeug-Verordnung of 21. June 2002 (BGBl. I S. 2214), last revision through Article 3 of the regulation from 20. Dezember 2010 (BGBl. I S. 2194)

EU, Directive 2000/53/EC of the European Parliament and of the Council of 18 September 2000 on end-of life vehicles—Commission Statements

Kaerger W (2010) Netzwerke der Demontagebetriebe—Fahrzeugrücknahmesystem und optimierter IT-gestützter Ersatzteilhandel. In: Recycling und Rohstoffe—Band 3, Karl J. Thome' Kozmiensky, Daniel Goldmann, Neuruppin: TK Verlag Karl Thome' Kozmiensky

KrW-AbfG (1994) Kreislaufwirtschafts- und Abfallgesetz, Germany, 27. September

Ming Ch (2005) End-of-life vehicle recovery in China: consideration and innovation following the EU ELV directive. J Miner Metals Mater Soc JOM 57(10):20–26

Ming Ch (2009) End-of-life vehicle recovery in China: now and the future. J Miner Metals Mater Soc JOM 61(3):44–51
Pehlken A, Müller DH (2009) Using information of the separation process of recycling scrap tires for process modelling. Resour Conserv Recycl 54(2):140–148
Pehlken A, Decker A, Thoben K.-D (2010) Life cycle management of secondary resources—sustainable resource for new products. In: 7th international conference on product lifecycle management PLM 2010, Bremen, Germany
Rueth E.(2011) Kleiner Bruchteil, Recycling Magazin 16, pp 16, 17
VDI Guideline 4080; Recycling of cars—Quality of recycled car parts; July 2009; VDI-HandbuchUmwelttechnik
VDI Guideline 4081; Recycling of cars—EDP management of recycled car parts; March 2004; VDI-HandbuchUmwelttechnik
VDI Guideline 4082; Recycling of cars—Draining and preparation of vehicles for disassembly; June 2006; VDI-Handbuch Umwelttechnik
Wang J, Ming Ch (2011) Recycling of electronic control units form end-of-life vehicles in China. J Miner Metals Mater Soc JOM 61(8):42–47
Xiang W, Ming C (2010) Implementing extended producer responsibility: vehicle remanufacturing in China. J Cleaner Prod. doi:10.1016/j.clepro.2010.11.016
Zhang T, Chu J, Wang X, Liu X, Cui P (2011) Development pattern and enhancing system of automotive components remanufacturing industry in China. Resour Conserv Recycl 55:613–622

Sustainability Issues Affecting the Successful Management and Recycling of End-of-Life Vehicles in Canada and the United States

Susan S. Sawyer-Beaulieu, Jacqueline A. Stagner and Edwin K. L. Tam

Abstract As the manufacturing and operation of vehicles become increasingly efficient, the environmental impacts of vehicles at the end-of-life phase become more significant. However, the effective recovery of recyclable parts and materials (particularly plastics) from end-of-life vehicles (ELVs) is fraught with challenges. For example, limited market demand for particular part and material types, dismantling difficulties (e.g., rusted part fasteners; welded parts assemblages), and non-uniformity of legislated controls and/or restrictions will influence the successful recovery and recycling of dismantled parts and materials. Automotive material variety and complexity combined with the limited effectiveness of processing technologies for liberating and separating automotive materials (plastics in particular) into sufficiently pure and recyclable material streams tend to limit materials recovery and recycling to principally automotive ferrous and non-ferrous metals. This chapter presents an overview and conceptual analysis of vehicle end-of-life issues to develop strategies and implement actions that can decrease the lifecycle impact of automobiles in their last and perhaps least understood stage.

Keywords End-of-life vehicles · Dismantling · Recycling · Plastics

S. S. Sawyer-Beaulieu (✉) · J. A. Stagner · E. K. L. Tam
Department of Civil and Environmental Engineering, University of Windsor,
401 Sunset Avenue, Windsor, ON N9B 3P4, Canada
e-mail: susansb@uwindsor.ca

J. A. Stagner
e-mail: stagner@uwindsor.ca

E. K. L. Tam
e-mail: edwintam@uwindsor.ca

P. Golinska (ed.), *Environmental Issues in Automotive Industry*,
EcoProduction, DOI: 10.1007/978-3-642-23837-6_10,

1 Introduction

Automobiles are an integral part of today's society; however, there is growing concern over the impact that these products have throughout their life cycle. Automobiles in many ways represent an extreme form of *complex consumer products*, such as computers, that pose significant challenges in how to assess their benefits and impacts on the environment and society. Although the use and production phases of a vehicle's life cycle are the largest contributors to the negative life cycle impacts of automobiles (Puri et al. 2009), the end-of-life phase must also be considered in order to have a complete understanding of the total life cycle of these products. The aim of this chapter is to summarize end-of-life vehicle processes, legislation, challenges, and to predict future issues in order to provide this more complete understanding of the end-of-life vehicle phase. Finally, by predicting how emerging technologies may affect end-of-life vehicle management, this can provide valuable insights and information to vehicle manufacturers and recyclers to aid in the decision-making process.

2 End-of-Life Vehicle Management in Canada and the United States

Assuming 6 % of all registered roadway vehicles in use are retired annually and 6 % of all retired vehicles are abandoned (Staudinger and Keoleian 2001), an estimated 13 million vehicles are permanently retired and recycled in Canada and the United States annually (Sawyer-Beaulieu 2009). These end-of-life vehicles (ELVs) are managed within a network consisting primarily of dismantlers (salvage yards and junk yards, included), crushers, and shredders, and metal manufacturers.

ELV dismantling and shredding practices and post-shredder recovery/treatment processes will vary from region to region, as influenced by:

- regulatory constraints (federal, provincial/state, municipal);
- market supply and demand for ELVs and used car parts;
- market value of the particular parts recovered;
- supply and demand of ELV hulks as shredder feedstock;
- shredder feed material specifications (i.e., acceptable versus non-acceptable materials) and quality control (i.e., inspection, sampling, testing of materials destined for shredding);
- shredder through-put capacity;
- downstream shredded material processing methods/technologies;
- shredded metal product quality control;
- supply and demand of shredded ferrous metals as alternative melting units for steel mills and foundries;
- foundry and steel mill feedstock specifications and quality control;
- shredder residue (SR) processing/management options (Sawyer-Beaulieu 2009).

2.1 ELV Dismantling Practices

ELV dismantling business models will vary, but, in general, dismantling businesses are typically operated as "full-service" facilities, "self-service" facilities, or as a combination of full and self service, frequently referred to as "hybrid" facilities (Haddad 2008; Sawyer-Beaulieu 2009). Full-service facilities dismantle the ELVs they receive using in-house personnel, recover and inventory the resalable parts, as well as inspect, test and clean the parts as may be required prior to their sale. In self-service facilities (commonly called "UPIC" or "U-Pull-It" facilities), ELVs are placed into a yard where customers may come and pull the parts themselves using their own tools, and buy them at a reduced price (Sawyer-Beaulieu 2009).

When a vehicle reaches its end-of-life, it may be retired as a consequence of old age and/or poor mechanical and/or physical condition, rendering the vehicles incapable of operation on roads or highways (i.e. unable to pass safety certification). A vehicle may also be retired as a 'write-off' as a result of severe damage (by collision, impact, fire, or flood) or theft and dismantling (Sawyer-Beaulieu 2009).

Vehicles that enter the dismantling process may be obtained from a number of sources, including insurance companies, auctions, dealers and the public (ARC 2011b). The vehicles are typically inspected and evaluated by the dismantlers according to their make, model, model year, physical condition, and by the value and demand for particular automotive parts (Sawyer-Beaulieu 2009). They are consequently classified and managed as either "high salvage" (late-model, typical) vehicles or "low salvage" (old-age/early-model, typical) vehicles after entering the facility (refer to Fig. 1). The high salvage vehicles are typically late-model, accident/collision vehicles retired as vehicle write-offs (also referred to as total loss vehicles or TLVs) (Sawyer-Beaulieu 2009). High salvage-value parts are identified and their respective parts information and vehicle administration data is entered into computer-based parts inventory management systems (Fletcher 2011; Sawyer-Beaulieu 2009). Fluids and hazardous parts and materials are recovered and directed for reuse, recycling, energy recovery, and/or disposal.

Recovered fluids typically include refrigerants, antifreeze, gasoline, windshield washer fluid, lubricants—engine oil, transmission oil, differential fluid, brake-line fluid and power steering fluid—and, to a lesser extent, shock absorber fluid (Hoeher 2009a, b; Hoeher and Michael 2010; Sawyer-Beaulieu 2009). Lubricants may be shipped offsite for recycling using a licensed waste hauler, or alternatively, may be used by the dismantlers in on-site oil-fired space heaters for comfort heating. Recovered refrigerants, antifreeze, gasoline and/or windshield washer fluid may be reused on-site by the dismantlers, sold to customers for off-site reuse, or shipped offsite for recycling using a licensed waste hauler (Hoeher 2009a, b; Hoeher and Michael 2010; Sawyer-Beaulieu 2009).

Hazardous or environmentally sensitive parts and materials removed from ELVs by dismantlers typically include batteries, un-deployed air bags, tires, catalytic convertors, fuel tanks, mercury switches, and lead wheel weights

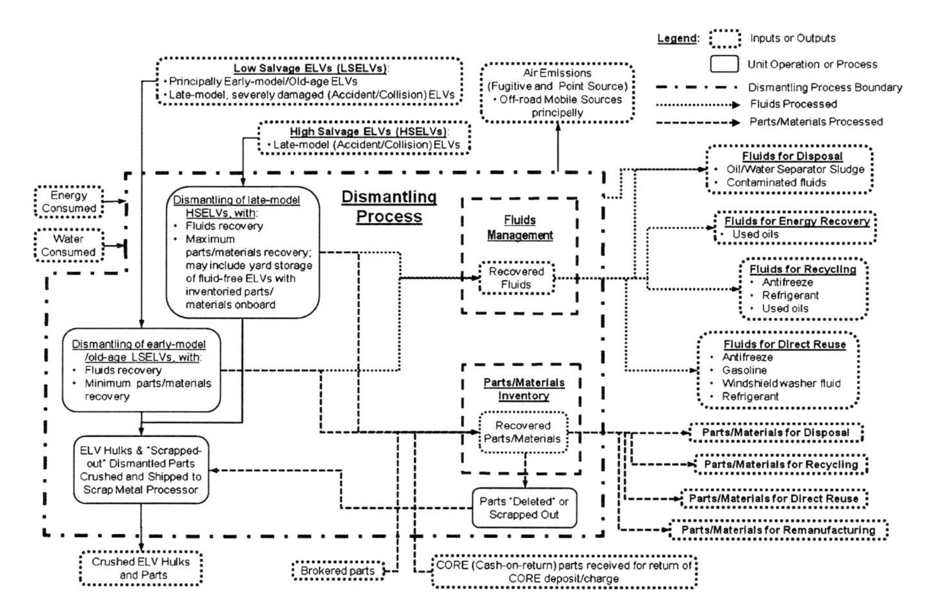

Fig. 1 Simplified dismantling process flow diagram (adapted from Sawyer-Beaulieu 2009)

(Keoleian et al. 1997; Staudinger and Keoleian 2001; Sawyer-Beaulieu 2009). Recovered batteries may be sold for reuse or directed for recycling. Un-deployed airbags will be either: (1) removed for reuse in jurisdictions that permit this, (2) deployed and left in the vehicles, or (3) removed, deployed and sent with the ELV hulks for shredding (Automotive Recyclers Association (ARA) 2011a; Canadian Council of Motor Transport Administrators (CCMTA) 2004; Sawyer-Beaulieu 2009). The recovery of automotive mercury-containing switches (i.e. hood/trunk convenience lights, ABS sensors) is largely performed under voluntary switch removal programs, such as the voluntary Switch Out Program coordinated by Summerhill Impact (former Clean Air Foundation) in Canada (Summerhill Impact 2011b), or National Vehicle Mercury Switch Recovery Program (NVMSRP) in the United States (End of Life Vehicle Solutions Corporation (ELVS) 2011). In some states, however, mercury switch recovery is required by law (i.e. Arkansas, Illinois, Iowa, Massachusetts, New Jersey, Rhode Island, Utah, Maryland, Indiana, North Carolina, South Carolina) (ELVS 2011). Tires are typically considered unacceptable shredder feed materials; they are usually removed by the dismantlers and either sold for reuse or sent for recycling (Staudinger and Keoleian 2001; Sawyer-Beaulieu 2009).

Parts removal and storage practices used by the dismantlers vary. Based on their assessment of the "principal" high salvage-value parts targeted for recovery and sale as reusable parts, some dismantlers remove these high value parts first, then place the "leftover" ELVs into inventory yards where inventoried parts are stored "on-board" the ELVs themselves for a certain period of time (Keoleian et al. 1997;

Staudinger and Keoleian 2001; Sawyer-Beaulieu 2009). This process allows the dismantlers access to other salvageable, but less popular parts, that are removed from the ELVs only after the higher value parts have been sold. Other dismantlers will strip any and all reusable parts identified for salvage, store only these parts, and not maintain yard storage of ELVs with on-board inventoried parts (Keoleian et al. 1997; Staudinger and Keoleian 2001; Sawyer-Beaulieu 2009). If the dismantlers do not have a particular part a customer is looking for, they may provide a "brokered part", a part brought in from another dismantler who has the part in inventory (Sawyer-Beaulieu 2009). Although dismantling is largely a manual process, power tools are used in preference to manual hand tools, wherever practical, and dismantlers often employ mechanized, semi-destructive dismantling techniques such as cutting, in which parts of negligible or lower value will be sacrificed to permit access to high value parts or assemblages (Sawyer-Beaulieu 2009).

Computer-based parts inventories are typically maintained and used to sell parts and to facilitate in deciding what to dismantle (Keoleian et al. 1997; Staudinger and Keoleian 2001; Recycling Council of Ontario (RCO) 1999; Fletcher 2011). Parts dismantled for reuse are each assigned an industry-wide interchange number that identifies which vehicle make, model and type it fits. The parts are then labeled with a bar code or inventory number, which is also entered into the computerized inventory management systems facilitating the tracking and locating of the parts in the dismantlers parts inventory or warehouse (Fletcher 2011). The computerized parts inventory management systems are typically interconnected through parts locator networks, connecting the inventory data of hundreds of auto recyclers across Canada and/or the United States. This permits an auto recycler to locate a part for a customer if the part is not available in the recycler's parts inventory (Fletcher 2011).

Prior to selling the parts to customers, dismantled parts are typically cleaned to remove dirt, oil and grease. To conserve water and reduce the amount of waste fluids generated, dismantling facilities commonly use closed-circuit parts washing systems: wash water is treated and reused within the system. Waste water generated as a consequence of water used in the dismantling process—typically oil/water separator sludge produced in a parts washing system—will be shipped by a contracted licensed waste hauler for off-site disposal (Sawyer-Beaulieu 2009).

Salvageable parts that are removed from the ELVs and determined to be unsuitable for sale as a reusable part, but are refurbishable, will commonly be sold by the dismantlers to parts remanufacturers (Staudinger and Keoleian 2001; Sawyer-Beaulieu 2009). Remanufacturable parts are generally referred to as "cores", analogous to an "apple core". An engine assembly, for example, that is tested and determined to be unsuitable for direct reuse may be stripped of reusable parts, leaving a "core" which itself may have value as a remanufacturable part (Sawyer-Beaulieu 2009). Parts commonly sold for remanufacturing include engines, starters, AC compressors, water pumps, carburetors, calipers, power steering pumps, carrier assemblies, windshield wiper motors, electronic control units (ECU), alternators, transmissions, axle assemblies and transfer cases (Johnson and Wang 2002; Sawyer-Beaulieu 2009).

Dismantlers will apply "Cash-On-Return" or CORE charges on certain part types (Sawyer-Beaulieu 2009). A CORE part is a part that may be received from a customer for return of a CORE deposit or charge. A "CORE charge" is a refundable deposit for the value of the CORE part that is paid at the time a "new" used part is purchased. The CORE part may be traded in for the credit of a portion of the price of the "new" used part being purchased (Sawyer-Beaulieu 2009). For example, instead of paying full price for a new part, such as an alternator, an old alternator can be submitted as a CORE and consequently reduce the price that the customer would have to pay for a "new" used alternator. CORE parts received by a dismantler will sometimes be sold as parts for reuse, but are most commonly sold with parts for remanufacturing, or directed for recycling with ELV hulks (Sawyer-Beaulieu 2009).

To facilitate the removal of the high salvage-value parts that the dismantlers target for recovery, other parts of little or no value may have to be removed first to make the desired parts accessible. Typically, these no-value parts are returned to the stripped vehicle and sent for shredding with other ELV hulks. Some stripped part types may not be returned to the ELVs, but will be shipped in segregated loads for shredding and metals recycling, e.g., steel or aluminum wheels (Keoleian et al. 1997; Staudinger and Keoleian 2001; Sawyer-Beaulieu 2009).

Periodically, dismantlers may perform an inventory clean-up. Dead or over-stock parts inventory is removed, or "scrapped-out", and sent for shredding with the ELV hulks (Staudinger and Keoleian 2001; Sawyer-Beaulieu 2009). ELVs that are to be scrapped-out and have parts inventoried on-board are reviewed for salvageable parts to be kept. Those parts are removed from the ELVs and the remaining hulks are sent for shredding (Sawyer-Beaulieu 2009).

Dismantlers commonly compact their leftover ELV hulks, along with scrapped-out parts, prior to shipping them to the shredders using either their own on-site car crushers or contracted portable car crushers (Keoleian et. al. 1997; Staudinger and Keoleian 2001; Sawyer-Beaulieu 2009). Compaction maximizes the number of ELV hulks that may be shipped at one time at the most economical cost while satisfying shipment height restrictions, where applicable. Some dismantlers may ship their ELV hulks and scrapped out parts without crushing them because of their close proximity to receiving shredding facilities and their low ELV processing throughputs, e.g., two or less ELVs per day (Sawyer-Beaulieu 2009).

2.2 ELV Shredding

As of 2006, there were about 220 auto shredding facilities in the U.S. (Taylor and Toto 2006) and approximately 20 in Canada. As illustrated in Fig. 2, stripped ELV hulks and scrapped out parts that are shipped to shredding facilities in Canada and the United States are typically processed through hammer mill shredders along with other metal-rich scrap materials, in particular end-of-life appliances (ELAs) or "white goods", and construction, renovation and demolition waste (Keoleian et al. 1997; Staudinger and Keoleian 2001; Sawyer-Beaulieu 2009).

The fragmented material discharged from a mill is further processed typically using air separation of the low density, non-metallic materials from the higher density, metal-rich fraction. The metal-rich fraction is subsequently processed by magnetic separation to separate the ferrous metals (cast iron, carbon steel) from the non-ferrous and non-magnetic metals (aluminum, copper, zinc, nickel, stainless steel, and lead). The shredded ferrous metal product is recycled as alternative steel mill feed stock (Keoleian et al. 1997; Staudinger and Keoleian 2001; Sawyer-Beaulieu 2009).

The predominantly non-ferrous, non-magnetic metal fraction, containing high grade stainless steels (SS), as well as some low density, non-metallic materials, usually requires further treatment, for example, using a combination of screening, air classification and eddy current separation methods, to improve metals recovery

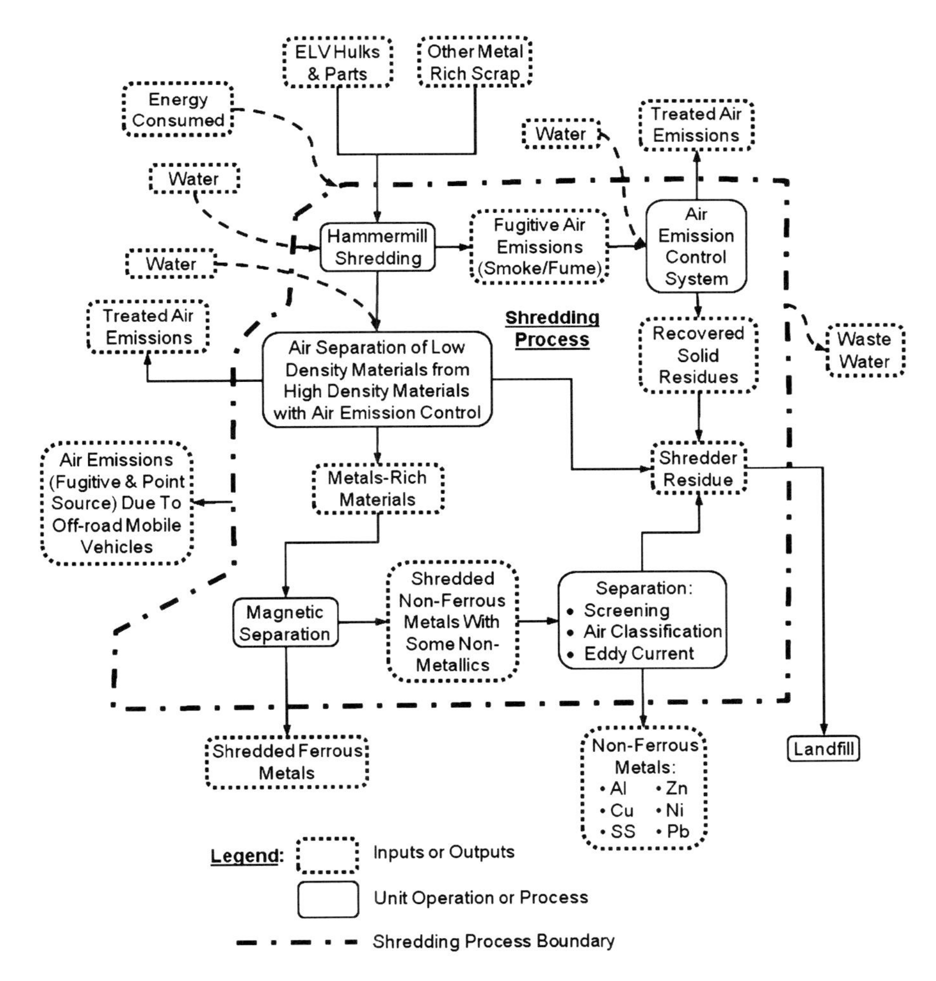

Fig. 2 Simplified shredding process flow diagram (adapted from Sawyer-Beaulieu 2009)

(Gesing et al. 1998; Sawyer-Beaulieu 2009). The resulting mostly non-ferrous metal product will typically be shipped to recycled metals processors for additional processing and treatment to separate the materials into individual metal fractions that are of sufficient purity for subsequent metal refining. The additional processing methods that may be used include, for example, screening, eddy current separation, heavy media separation, and air-fluidized sand-bed separation. The left-over, mostly non-metallic materials, i.e. the shredder residue (SR), is routinely disposed of by landfilling (Keoleian et al. 1997; Staudinger and Keoleian 2001; Sawyer-Beaulieu 2009).

Water will be strategically added into a shredding process in the mill and air separation/emission control systems in controlled quantities to control mill temperature, to prevent fire, reduce wear on mill parts and to help control fugitive air emissions generated by the process. The quantity of water that may be applied can vary from minimal quantities—for example just sufficient quantities to keep fires in check, i.e. "dry shredding"—to flooded conditions (i.e. "wet shredding") (Sawyer-Beaulieu 2009). The flooded conditions of wet shredding effectively prevents the generation and discharge of fugitive air emissions from the mill and, hence, avoids the requirement (and cost) for an air emission collection and control system. Wet shredding, however, results in mill discharge materials saturated with water requiring some sort of system for dewatering the materials and handling the waste water generated by the process. Further, the SR generated in a flooded shredding process is significantly heavier resulting in higher transportation and disposal costs (Sawyer-Beaulieu 2009).

Air emission control systems are typically required in shredding processes (except wet shredding) for collection and treatment of fugitive air emissions generated and discharged from a shredder mill, and in the air separation systems that remove the non-metallic materials from the heavier metals-rich stream (Institute of Scrap Recycling Industries (ISRI), Inc. 1998; Sawyer-Beaulieu 2009). The air emission control equipment commonly consists of at least an air cyclone separator for collecting larger particulates, and could include a wet scrubber for removing fine particulates, oil mists/fumes, etc. from the air stream (ISRI, Inc. 1998, Sawyer-Beaulieu 2009). Although not considered the best available technology (BAT), wet scrubbers are typically used in preference to air filtration systems for treating shredder air emission streams to avoid the risk of fire. Scrubber water is typically collected and recirculated, eliminating the need of waste water treatment and discharge (ISRI, Inc. 1998; Sawyer-Beaulieu 2009).

3 Regulation of ELV Management in Canada and the US

The ELV management industry is well established in North America and the processing technologies are generally understood, however the regulatory aspects of ELV management are not well-known/understood (Sawyer-Beaulieu 2009). Regulatory issues impacting ELV dismantling and shredding facilities can include:

- environmental site development licensing;
- facility/business operations licensing;
- business-related or operations-related compliance documentation and reporting;
- zoning bylaws restricting site use;
- management, control and permitting of environmental emissions (air emissions, noise, waste water and storm water discharges, hazardous waste disposal);
- environmental performance/compliance reporting.

The regulation of ELV management facilities in Canada and the United States is primarily focused on business and operating practices as opposed to the regulation of the retired vehicles themselves. The regulatory mechanisms that are applied include legal statutes, regulations, and bylaws, as well as voluntary mechanisms, such as best management practices (BMPs). The operations, activities and practices that are typically regulated and/or controlled in ELV management businesses include:

- emission of air contaminants, including noise;
- discharge of waste water (process and/or storm water);
- waste generation and disposal;
- site use and materials storage.

In addition, these facilities typically require business licensing (under provincial/state legislation and/or municipal bylaws), which permits them to carry out dismantling and recycling of ELVs. Municipal bylaws governing the licensing of ELV dismantling and recycling commonly stipulate site-use conditions or restrictions such as materials storage restrictions or site accessibility conditions (Sawyer-Beaulieu 2009).

ELV management companies in Canada and the United States are subject to business and environmental regulations, however the specific legislative criteria and mechanisms vary somewhat from jurisdiction to jurisdiction. For example, the use of used crankcase oils as fuel in used-oil fired space heaters, for comfort heating, is regulated in Canada and the United States.

In the Unites States, a permit or license is not required for the burning of used oil in used-oil fired space heaters as long as (1) the used oil is generated on site or collected from a "do-it-yourselfer" used oil collection center (DIYs), (2) the space heater used has a maximum capacity of 0.5 million Btu per hour or less, and (3) the combustion gases from the heater are vented to the ambient air (United States Environmental Protection Agency (USEPA) 1994; United States 2010). In Canada, the regulated use of used oil as a fuel varies slightly from province to province. In some provinces, facilities have to be permitted or licensed (e.g. Certificate of Approval) to use used-oil fired space heaters, and only equipment approved by the Canadian Standards Association (CSA) and Underwriters' Laboratories of Canada (ULC) may be used. In some provinces small used-oil generating businesses are exempted from the permits required by larger sites if they register (allowing them to use used-oils as fuel), and they do not exceed specified maximum use rates (e.g. 15 liters per hour per premises) and the used oil-fired heater conforms to specified equipment standards (e.g. CSA) (Environment Canada 2005, 2011; New Brunswick 2002).

The management of end-of-life vehicles (ELVs) in Canada and the United States is largely a market driven industry, with used parts and scrap metal prices driving high recycling rates (ARC 2011b). With the exception of British Columbia, there is no jurisdiction in Canada and the United States where ELV management activities are legislated. British Columbia's Vehicle Dismantling and Recycling Industry Environmental Planning Regulation (VDRIEPR) requires a dismantler processing 5 or more ELVs per calendar year to establish, register, follow and maintain an environmental management plan (EMP) for the ELVs they process (British Columbia 2007). The EMP must describe how prescribed wastes (liquids, refrigerants, batteries, mercury switches and tires) are removed, stored, treated, recycled and/or disposed. It must also define management processes for minimizing or eliminating the discharge of waste to the environment (British Columbia 2007). The VDRIEPR also outlines auditing and reporting requirements for vehicle dismantlers. Every five years, a dismantler's EMP must be reviewed, amended and approved by a qualified professional (as defined under the regulation). Every two years, each vehicle dismantler must be audited by a qualified professional and an audit report prepared confirming how the prescribed wastes were managed, if they were managed in accordance with the facility's EMP and how effective the facility's management processes are in minimizing or eliminating the discharge of wastes to the environment. (British Columbia 2007; British Columbia Ministry of Environment (BCMOE) 2008).

A number of industry standards or codes of practice are being used in the dismantling and metals recycling industries that help to regularize ELV recycling practices. Recycled parts grading guidelines, standardized part type definitions and descriptions, and part damage location and identification codes, developed by the Automotive Recyclers Association (ARA) (ARA 2006, 2011b) have been adopted on an international scale and are built into auto recyclers' inventory systems (Fletcher 2011). These parts identification and grading codes have been established for automotive body panel parts and mechanical parts and facilitate recycled parts quality control.

The Canadian Council of Motor Transport Administrators (CCMTA) guidelines for use of recycled original equipment air bags outline procedures to safely re-use "recycled", i.e. non-deployed, OEM airbags, in jurisdictions where it is permitted (ARC 2011a; Canadian Council of Motor Transport Administrators (CCMTA) 2004).

In 2008, the National Code of Practice was established for automotive recyclers participating in Canada's National Vehicle Scrappage (Retire Your Ride, http://www.retireyourride.ca/home.aspx) Program (Automotive Recyclers of Canada (ARC) 2008; Summerhill Impact 2011a). Created by the Automobile Recyclers of Canada (ARC) for Environment Canada, the code required all automotive recyclers enrolled in the program to comply with applicable legal requirements, e.g. de-registration of vehicle identification numbers (VINs), as well as with environmental management practices specified under the Code, including:

1. recovery, storage, transportation, manifesting, disposal and/or record keeping requirements for waste fluids (including refrigerants), lead acid batteries, mercury-containing switches, lead wheel weights, and other hazardous materials or components;
2. training in practical best management practices (BMPs);
3. vehicle processing area requirements (i.e. vehicle receiving, dismantling, hulk storage, crushing, "wet parts" storage, hazardous fluids storage); and
4. facility audit requirements to ensure compliance with the code.

When the National Vehicle Scrappage Program terminated March 31, 2011, 137,783 vehicles had been permanently retired and processed by 335 participating recyclers across Canada (Summerhill Impact 2011b).

Although extended producer responsibility (EPR), or product stewardship, practices have not been legislated for ELVs managed in Canada and the United States, EPR based initiatives have been launched for ELVs and/or ELV derived materials, or are under consideration. Canada's Switch Out Program and the NVMSRP in the United States (ELVS 2011; Summerhill Impact 2011b), for example, are EPR-based partnerships between automotive manufacturers, automotive recyclers/dismantlers, scrap metal recyclers, and steel manufacturers, committed to the recovery and recycling of automotive mercury-containing switches from ELVs and hence reducing the release of mercury to the environment (Summerhill Impact 2011b; USEPA 2006). Quebec currently has EPR programs for mercury switches, used oil and used oil filters with plans to add programs for used tires and automotive electronics in the near future (ARC 2011b; Quebec 2008).

In 2009, the Ontario Ministry of the Environment outlined proposed EPR-based changes to Ontario's waste diversion framework, which would include the banning of ELVs and ELV-derived materials from landfill disposal, with proposed five-year material-specific collection and diversion targets (Ontario Ministry of the Environment (OMOE) 2009). The proposed EPR-based waste diversion system would make individual producers (manufacturers, brand owners, or first importers of products or packaging made with a designated material) responsible for meeting diversion outcomes (OMOE 2009).

More recently, EPR-based, environmental management system (EMS) standards have been proposed for ELV management in Ontario (Ontario Automotive Recyclers Association (OARA) 2011), as well as for ELV management across Canada (ARC 2011b). These proposed provincial and national standards are based on the use of the National Code of Practice that was established for automotive recyclers that participated in Canada's National Vehicle Scrappage Program (ARC 2008). The proposed ELV EMS standards would require all ELV processors to be licensed, authorized or certified under provincial law, and subject to a common decommissioning standard (codified in provincial law) to minimize environmental discharges and ensure proper treatment of substances of concern. It is suggested that the proposed standards would help to level the playing field for businesses in the ELV management industry, while maintaining the market-driven, competitive structure of the automotive parts and materials recycling industry (ARC 2008; OARA 2011).

4 Closed-Loop Recycling and Plastics in End-of-Life Vehicles

As vehicle manufacturers strive to improve the fuel economy of vehicles, reducing vehicle weight has become increasingly important. Many plastics are economical, easy-to-manufacture materials that can reduce the weight of components that have historically been produced from materials such as metals. Figure 3 shows how the weight of plastic within vehicles has increased since 1985, with plastics contributing the second largest weight percentage of material, behind only ferrous metals (Ferrão and Amaral 2006a; The Society of Motor Manufacturers and Traders Limited 2011). However, plastics are difficult to recover when a vehicle reaches the end-of-life phase. Plastics can be used in vehicle parts or locations that are difficult to access. This situation is often made more complex when plastics are joined or fused with other plastics, making them difficult to economically separate. Challenges in recovery therefore hinder actual recycling.

Recycling can be divided into four main categories; closed-loop recycling, downgrading, chemical or feedstock recycling, and energy recovery (Hopewell et al. 2009). Closed-loop recycling is what most people envision when they hear the term "recycling". In closed-loop recycling, the materials from an end-of-life product are used to produce a new product with equivalent properties (Hopewell et al. 2009; Chilton et al. 2010). Thus, the recycled materials have more value than if they were simply used to replace cheap filler materials (Palmer et al. 2009).

Recycling materials so that they can produce products with equivalent properties is an ideal situation. However, it is difficult to achieve equivalent properties in recycled materials as compared to virgin materials. Plastics age and this impacts

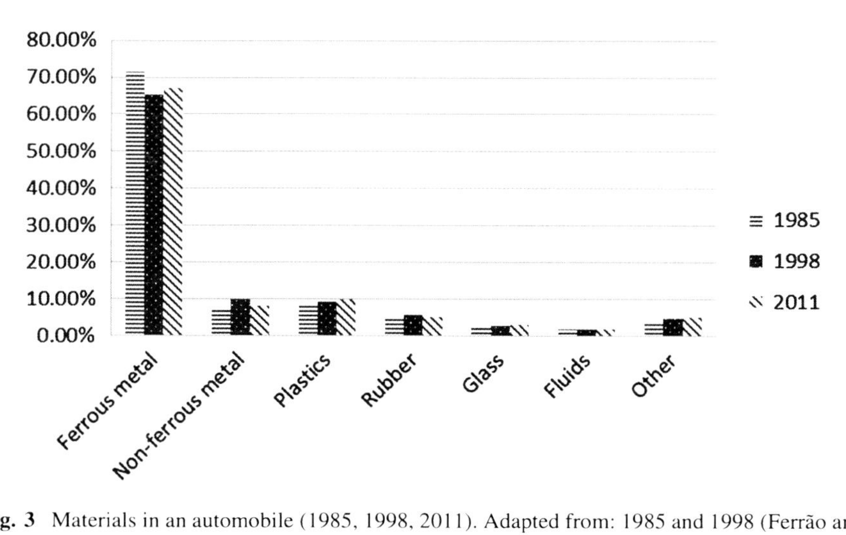

Fig. 3 Materials in an automobile (1985, 1998, 2011). Adapted from: 1985 and 1998 (Ferrão and Amaral 2006a), 2011 (The Society of Motor Manufacturers and Traders Limited 2011)

Table 1 Elongation at break for ABS and HDPE after extrusion and after ageing and extrusion cycles

Material	ABS[a]	HDPE[a]
Elongation at break (after 1st extrusion)	9 %	750 %
Ageing conditions	72 h at 90 °C	48 h at 110 °C
Number of ageing and extrusion cycles	6	10
Elongation at break (after ageing and extrusion cycles)	3 %	770 %

[a] *Sources* ABS (Boldizar and Möller 2003), HDPE (Boldizar et al. 2000)

the long-term behavior of products made from these materials (Struik 1977). Furthermore, the recycling of plastics involves not only aged plastics, but also plastics that have been extruded multiple times.

Studies on ABS, PP, and HDPE have tried to simulate the effects of aging and extrusion that occur during the recycling of plastic materials (Boldizar and Muller 2003; Boldizar et al. 2000; Strömberg and Karlsson 2009) (see Table 1). Table 1 shows that the combined effects of aging and recycling on the mechanical properties of various plastics is not consistent. Furthermore, industrially recycled polymers have shown poorer mechanical properties than those measured during simulated recycling conditions (Strömberg and Karlsson 2009).

Furthermore, the contamination of materials (i.e. from different plastics, fillers, pigments, additives) produces plastics with degraded properties. Material contamination can result in plastics with low tensile strength, mixed colour, and decreased transparency (Astrup et al. 2009). Thus, recycled plastics are often blended with virgin plastics in order to ensure that the resulting material has desirable properties (Astrup et al. 2009).

Although there are technical challenges to closed-loop recycling of plastics, studies have shown that from a life-cycle assessment viewpoint, closed-loop recycling of polyethylene terephthalate (PET) bottles results in an overall reduction in emissions of key pollutants studied, as well as a reduction in the overall environmental impact, as compared to recovering the PET waste and burning it in an energy-from-waste facility (Chilton et al. 2010); however, the economic implications were not studied. Studies that have looked at the economics of recycling ELV parts have shown that it may not be economically justifiable to recycle these parts, and is dependent on the marketplace demand for recycled materials (Bellmann and Khare 2000).

5 Other End-of-Life Recycling Options to Improve Sustainability

Downgrading refers to plastics that are of lower quality than virgin plastic. This may be due to contamination of different plastics, fillers, additives, colors, etc. Certain properties that can be affected are the tensile strength, transparency, and colour of the plastics (Astrup et al. 2009). These plastics are often used as fillers in products such as pallets, fences, and garden furniture.

Chemical or feedstock recovery has the advantage of producing valuable and useful products that can be used to make new petrochemicals or plastics (Al-Salem et al. 2009). Methods that enable chemical or feedstock recovery include pyrolysis, gasification, and solvolysis (Zia et al. 2007; Sasse and Emig 1998). In pyrolysis, a material is heated to a controlled temperature, in the absence of oxygen, causing volatile organic materials to decompose into gases and liquids. In the case of automotive parts, the volatile organic materials would be the plastics and rubbers used in the components. The other materials (metals, glasses, etc.) remain unchanged and can be separated from the organic matter (de Marco et al. 2007). The organic gases and liquids obtained through pyrolysis can be used as fuels or as a source for organic chemicals (Buekens and Huang 1998). In this way, the basic elements from ELV plastic components are reused. A benefit of this method of reusing polymeric materials is that the shredder residue (SR), which is produced when ELVs are shredded, can be pyrolised and the different types of plastics do not need to be separated from each other. In the gasification process, a material is also heated to a controlled temperature; however, it is in the presence of air (Al-Salem et al. 2009).

Solvolysis is a term used to describe the depolymerization process in which the original monomers are produced from the materials. Various types of solvolysis include hydrolysis, glycolysis, and methanolysis (Sasse and Emig 1998), in which polymer chains are cleaved by reagents such as water, glycols, or methanol, respectively (Sinha et al. 2010). It should be noted that the polymers used in these processes should be sorted and precleaned.

In energy recovery, waste plastics are incinerated and the heat is recovered. This heat may be used to heat a space or can be used to generate power (Eriksson et al. 2005). Also, using SR as fuel in a blast furnace has shown promising results (Mirabile et al. 2002). An advantage of energy recovery methods is that separation of the various plastics is not necessary. However, SR is non-homogeneous and the energy content can vary.

Polymer composites further complicate the recyclability of automotive materials. Composites are materials composed of two or more constituent phases (Vidal et al. 2009). Polymer composites have a polymeric matrix and a reinforcement or filler phase. In the case of composites, a combination of the various types of recycling may be utilized. Mechanical recycling of composites breaks down the composites by mechanical means and then separates resin-rich powder products from fiber-rich fibrous products. These materials can be reintroduced into new composites as fillers and reinforcements; however, these products are low-value applications. Another method, fiber reclamation, is particularly useful for carbon fiber reinforced composites. In fiber reclamation, thermal or chemical processes are used to break down the polymeric matrix and the fibers are released and collected. Chemical and/or energy recovery of the matrix is possible (Towle 2007). Furthermore, composites can be produced using reclaimed carbon fibers. The recycled carbon fibers have exhibited a high retention of mechanical properties (Towle 2007).

Bio-based, biodegradable materials offer an additional end-of-life treatment method: composting. Kim et al. (2008) have reported that composting of an automobile component made from a fiber reinforced biocomposite is a more favorable waste management scenario than landfilling the materials.

6 Sorting and Separating Plastics

The recycling process creates shredder residue which in turn contains the polymeric materials to be recycled. These various polymers may be used to generate energy or fuels, in which case, it is not necessary to separate. However, energy recovery scenarios have not been shown to be promising in meeting EU recycling targets of 85 %, due to stringent regulatory requirements (Ferrão et al. 2006c).

If the polymers are to be recycled into polymers to make new products, they must first be separated, cleaned, and free of contaminants. Usually, separation is thought of as parsing distinct materials from one another, such as in "blue box" or curbside collection programs of recyclables. Simple recyclable items, such as water bottles, newspapers, etc., may be co-mingled, but inherently are not attached to one another. Automotive plastic parts are unlikely to be so simply configured. They often combine multiple plastic types and various fastening or joining methods. The recovery of automotive plastics therefore means that separation must be thought of in two aspects: (1) liberation; and (2) actual separation of materials from one another.

In terms of the latter, separation, there are many different methods of separating materials. Dalrymple et al. (2007) have listed screeners; air and water classifiers; density, electrostatic, and magnetic separators; and flotation systems as examples. It has been reported (Jody and Daniels 2006) that mechanical separation techniques are able to separate inorganic fines and residual metals from shredder residue, producing a polymer concentrate. Polymer separation techniques (i.e.: froth flotation) are then able to separate and recover polyolefins and engineered plastics (i.e.: ABS) from the polymer concentrate. A mixed-rubber fraction can also be separated and recovered using dry and wet processes (Jody and Daniels 2006). But to have separation, it is often assumed that different materials are no long *co-joined* to one another.

In terms of the former, liberation, size reduction through separation is often the most common operation. Breaking items down into smaller pieces should also liberate one material from another (Loehr and Melchiorre 1996) because bonds between different materials should be broken, be they adhesive in nature, welded, or bound by fasteners. The study of liberation as it applies to complex consumer goods has been relatively limited. Jekel and Tam (2007) showed that accepted ways of modeling how plastics could break down for eventual separation were not readily applicable or had limited value. The advantage of such a classification is that if plastics breakage could be predicted, designers and engineers could then exploit this to created plastic parts that are ultimately more recyclable. A key challenge to increasing the recovery of waste automotive plastics is how one plastic type can be liberated from another prior to applying various separation technologies.

To enhance the liberation of one material from another, some work has been performed using cryogenics to study the liberation of plastics from metals or other non-plastics during the recycling of ELVs (Gente et al. 2004; Dom et al. 1997). There is the potential of using differences in the behavior of plastics at cryogenic temperatures to aid in the separation of various plastics during ELV recycling. In addition, Barsha and Tam (2009) argue that the actual unit operation used in size reduction can play a far more significant role in whether or not plastics can be liberated from one another. Thus, results from one shredding operation in one facility may not be equal to results that are obtained in another facility. A successful design-for-environment effort to consider recovery in advance will require detailed knowledge of how a part or material will be handled at its end-of-life.

Sorting and separating plastics from shredder residue is not currently economical. In order to meet the 2006 EU requirements for ELVs (European Union 2000), Ferrão and Amaral (2006a) have shown that more thorough dismantling of plastic components from vehicles was sufficient; however, to meet the 2015 requirements, it has been suggested (Ferrão and Amaral 2006a) that separation technologies must be upgraded. Nevertheless, from an economics standpoint, the products liberated from the separation processes must have value in the marketplace.

7 Effects of Emerging Technologies on the Recyclability of End-of-Life Vehicles

Given the challenges of recycling polymeric materials at the automotive end-of-life phase, how will emerging technologies in automobiles affect the use of polymers? Reducing noise and vibrations to improve customers' driving experience, lightweighting of vehicles to reduce fuel consumption, and designing components that are more sustainable or environmentally preferable all affect the amount and types of polymers used in future vehicles.

Polymers, due to their viscoelastic nature, have been successfully used to damp noise and vibrations within vehicles (Rao 2003). Advances in technology now allow the cost effective mass production of multilayered laminate structures and spray-on dampeners (Rao 2003). This provides customers with the vehicle quality and performance that they expect from their vehicles. Nevertheless, these materials are quite difficult to separate at the end of a vehicle's useful life.

Bio-based polymers and composites are increasingly being used in vehicles (Suddell 2007). These materials are lightweight and can be used in many different applications within a vehicle (Kim et al. 2008). As well, they have the added benefit of being more environmentally preferable than traditional petroleum-based polymers (Kim et al. 2008). These materials also have the potential of being

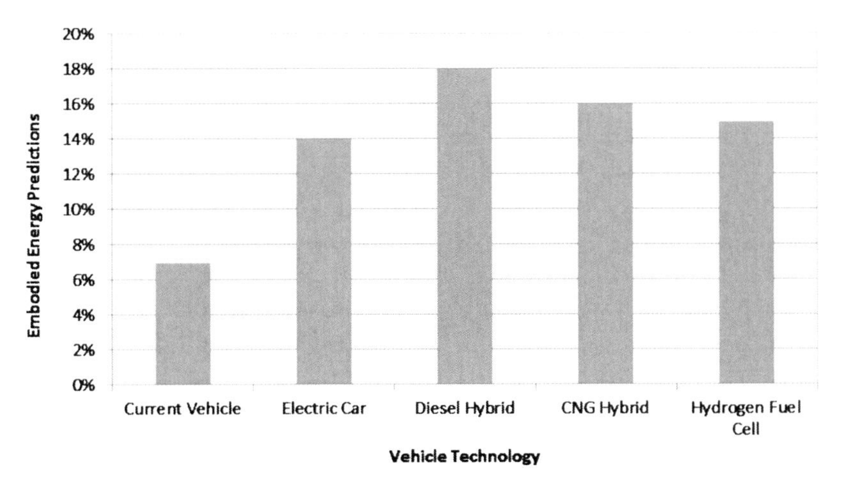

Fig. 4 Predictions of embodied energy (as a percentage of the life-cycle energy use) of vehicle technologies in the year 2020, based on a recycling rate of 95 % for metals and 50 % for plastics; adapted from Weiss et al. (2000)

collected and composted at the end-of-life phase. This can provide further environmental benefits from a lifecycle analysis perspective (Kijchavengkul and Auras 2008). Nevertheless, these bio-based materials add additional complexity to the number of different materials that must be sorted and separated during the recycling of a vehicle.

Emerging fuel technologies can also affect the lifecycle impacts of vehicles. Weiss et al. (2000) have shown that the embodied energy within the materials used in the production of vehicles can vary greatly, depending on the vehicle technologies (see Fig. 4). As vehicle technologies develop and new materials are used, or lightweight materials are used more extensively, the embodied energy within the vehicle materials changes. Furthermore, whether the materials used are virgin materials or recycled materials also have an effect. For the values displayed in Fig. 4, a recycling rate of 95 % for metals and 50 % for plastics has been assumed. These rates are not currently achieved in practice, but it has been assumed that they will be by the year 2020 (Weiss et al. 2000).

8 Considering the End-of-Life Phase for Vehicles During the Design Process

The method in which a component is manufactured or assembled can also affect the ease of recyclability of the vehicle from which it comes. Design for recycling (DfR) is a method in which end-of-life recyclability is considered during the design of a product. In the case of automobiles, Ferrão and Amaral (2006b) have shown that in order to meet increased recycling rates, dismantling, shredding, and post-shredding activities must occur. Consequently, automobile manufacturers are starting to adopt DfR principles during the process of designing their products (Gerrard and Kandlikar 2007).

As suggested by Ferrão and Amaral (2006a), there are two strategies available to increase the recycling rate of the constituent materials of shredder residue:

1. Decrease the amount of shredder residue created by more thorough dismantling of the complex products; and
2. Upgrade the sorting and separation technologies for shredder residue, and determine recycling possibilities for the materials collected through separation.

The first strategy can be aided by implementing design for disassembly (DfD) principles during the design process for new products. From a materials perspective, design for disassembly rules include minimizing the number of different materials used, using recyclable materials, and eliminating hazardous materials when designing products (Bogue 2007). To target the second strategy listed above, proponents of design for disassembly principles suggest that if multiple plastics must be used, their densities should differ by at least 0.3 (Bogue 2007). This enables the technique of separating plastics based on density to be more effective.

Using DfR software, it has been shown that dismantling tires, bumpers, glass, fuel tanks, and seats from vehicles can increase the recycling rate from 77.3 to 86.2 % and reduce the amount of materials that must be disposed by 42 % (Santini et al. 2010). This demonstrates the necessity of modifying current vehicle waste management techniques as well as designing vehicle assemblies to make their end-of-life dismantling more efficient and economical.

9 The Future of End-of-Life Vehicle Recycling

There are many challenges facing automobile manufacturers, dismantlers, and recyclers. Nevertheless, there is much that can be learned from other industries. For example, waste electrical and electronic equipment (WEEE), end-of-life aircraft, and end-of-life vehicles are all complex products which contain many different materials, including ferrous and non-ferrous metals, polymers, and composites. These materials must be managed when the products have reached their end-of-life.

Each sector has its own issues and concerns in regards to end-of-life recycling of plastics. Recyclers of WEEE must contend with many different consumer products, manufacturers, plastics, collection locations, and hazardous compounds (Schlummer et al. 2007; Schluep et al. 2009). By comparison, the collection of vehicles for recycling is better documented and reported (The Alliance of Automobile Manufacturers 2011), ELVs contain many different plastics, and there is a large volume of plastics in each vehicle. Finally, recycling of end-of-life aircraft is centralized and there are fewer different models of aircraft recycled (Michaels 2007; de Brito et al. 2007), as compared to WEEE and ELVs. However, predictions of future aircraft are that up to 50 % of the unladen weight in their primary structures will be composed of polymeric composites, mainly carbon fiber reinforced composites (Towle 2007). Composites are inherently difficult to recycle due to issues such as their complex composition, cross-linking of thermoset resins, and their combinations with other materials (Pimenta and Pinho 2011).

Much research has been performed in the various sectors to study how these products can be recycled more efficiently and economically (Kahhat et al. 2008; Schluep et al. 2009; Kim et al. 2004; Dalrymple et al. 2007; Towle 2007) and knowledge gained in one sector can benefit all sectors.

As discussed previously, emerging technologies may also add to the complexity of recovering ELV materials. Thus, DfR and DfD strategies must be embraced by vehicle manufacturers in order to account for all materials during the entire life cycle of their products. This design approach will ensure maximum recovery of materials as emerging technologies are introduced into new vehicles and vehicle complexities increase.

Finally, when life cycle assessments of end-of-life vehicles or components have been performed, there is general consensus that recovery of materials for recycling and/or energy recovery is the most environmentally preferable option (Puri et al. 2009; Duval and MacLean 2007), although waste-to-energy applications result in the consumption of resources that are predominantly non- renewable (e.g. petroleum-based plastics). There may not be one solution that best meets all of the challenges faced by ELV management, especially when predicting the handling of vehicles containing emerging technologies (e.g., electric vehicles, bio-based materials, laminated materials); however, a combination of technologies could provide more sustainable solutions.

Developing and implementing technologies for the recovery of ELV plastics prior to shredding could be simpler and of greater benefit than post-shredder ELV plastics recovery technologies. Rather than shredding the entire hulk with minimal prior hand disassembly, alternative "dismantling techniques or strategies" may be identified for recovering ELV parts/materials for recycling prior to shredding. For example, intermediate or limited separating processes (e.g., breakage, cutting, comminution) may be able to liberate additional items, which then may be processed by secondary or even tertiary processes (Sawyer-Beaulieu 2009).

Previous research on the economics of automobile dismantling under the 1990s North American recycling market (Johnson and Wang 2002; Spicer 1997) deemed disassembly of non-metallic components to be labor intensive and generally uneconomical. More recently, however, researchers are using industrial engineering systems approaches to model and assess the viability of "selective" or "targeted" parts dismantling and recovery scenarios, particularly for parts having high plastics content (e.g., automotive seat assemblies) (Barakat 2011), with the goal of reducing the amount of shredder residue requiring landfill disposal.

It will take a combination of government legislation, vehicle DfD and DfR advancements, technological advances in materials separation, and the implementation of more thorough dismantling procedures to make ELV recycling environmentally and economically sustainable.

References

Al-Salem SM, Lettieri P, Baeyens J (2009) Recycling and recovery routes of plastic solid waste,(PSW): a review. Waste Manag (Oxford) 29(10):2625–2643

Automotive Recyclers Association (ARA) (2011a) Air bags, the benefits of OEM non-deployed airbags. http://www.a-r-a.org/content.asp?contentid=526. Accessed 02 Sept 2011

ARA (2011b) ARA recycled parts standards and codes. http://www.a-r-amedia.org/a-r-a_org/ARA_standards_and_codes_full_final_cieca. pdf. Accessed 29 Aug 2011, 46 pp

ARA (2006) Part descriptions guidelines version 2.1—2006. http://www.a-r-a.org/files/SC_Parts_Description_10-10-06.pdf. Accessed 29 Aug 2011, 8 pp

Automotive Recyclers of Canada (ARC) (2008) National code of practice for automotive recyclers participating in the National vehicle scrappage program, for environment Canada, 36 pp (November)

ARC (2011a) Non-deployed OEM airbags. http://www.autorecyclers.ca/fileUploads/1263669760–Recycled_OE_Airbags.pdf. Accessed 29 Aug 2011, 3 pp

ARC (2011b) A national approach to the environmental management of end-of-life vehicles in Canada: Submission to the Canadian council of ministers of the environment. http://www.autorecyclers.ca/fileUploads/1313074351–National_ELV_EMS_approach.pdf, Accessed 29 Aug 2011, 21 pp (July)

Astrup T, Fruergaard T, Christensen TH (2009) Recycling of plastic: accounting of green-house gases and global warming contributions. Waste Manag Res 27(8):763–772

Barakat S (2011) Utilizing a systems engineering approach to evaluate end-of-life opportunities for complex automotive seat subassemblies. Master of Applied Science Thesis, University of Windsor, Windsor, Ontario, 140 pp

Barsha NAF, Tam EKL (2009) The potential effects of alternative comminution methods for enhancing recycling from plastic products. In: Paper 581, A&WMA 102nd Annual conference and exhibition, Detroit, Michigan, 13 pp (June)

Bellmann K, Khare A (2000) Economic issues in recycling end-of-life vehicles. Technovation 20(12):677–690

Bogue R (2007) Design for disassembly: a critical twenty-first century discipline. Assembly Autom 27(4):285–289

Boldizar A, Jansson A, Gevert T, Möller K (2000) Simulated recycling of post-consumer high-density polyethylene material. Polym Degrad Stab 68(3):317–319

Boldizar A, Möller K (2003) Degradation of ABS during repeated processing and accelerated ageing. Polym Degrad Stab 81(2):359–366

British Columbia (2007) Vehicle dismantling and recycling industry environmental planning regulation, B.C. Reg. 200/2007, consolidated regulations of British Columbia, 7 pp

British Columbia Ministry of Environment (BCMOE) (2008) Guidebook for the vehicle dismantling and recycling industry environmental planning regulation, 53 pp (July)

de Brito MP, van der Laan EA, Irion BD (2007) Extended producer responsibility in the aviation sector. ERIM report series research in management Erasmus Research Institute of Management. Research paper ERS-2007-025-LIS

Buekens AG, Huang H (1998) Catalytic plastics cracking for recovery of gasoline-range hydrocarbons from municipal plastic wastes. Resour Conserv Recycl 23(3):163–181

Canadian Council of Motor Transport Administrators (CCMTA) (2004) CCMTA guidelines for use of recycled original equipment air bags, D&V—May 2004, agenda item 17. http://www.ciia.com/newsletters/Guideline-Recycled-20040510-final-1.pdf. Accessed 29 Aug 2011, 23 pp

Chilton T, Burnley S, Nesaratnam S (2010) A life cycle assessment of the closed-loop recy-cling and thermal recovery of post-consumer PET. Resour Conserv Recycl 54(12):1241–1249

Dalrymple I, Wright N, Kellner R, Bains N, Geraghty K, Goosey M, Lightfoot L (2007) An integrated approach to electronic waste (WEEE) recycling. Circuit World 33(2):52–58

de Marco I, Caballero BM, Cabrero MA, Laresgoiti MF, Torres A, Chomón MJ (2007) Recycling of automobile shredder residues by means of pyrolysis. J Anal Appl Pyrol 79(1–2):403–408
Dom R, Shawhan G, Kuzio J, Spell A, Warren M (1997) Studies on optimizing the cryogenic process for recycling plated plastic to achieve improved quality of remolded parts. In: SAE international congress and exposition, Detroit, Michigan. Paper 970664
Duval D, MacLean HL (2007) The role of product information in automotive plastics re-cycling: a financial and life cycle assessment. J Clean Prod 15(11–12):1158–1168
End of Life Vehicle Solutions Corporation (ELVS) (2011) Mercury switches, ELVS website. http://www.elvsolutions.org/mercury_home.htm. Accessed 28 Aug 2011
Environment Canada (2005) Follow-up report on a PSL1 substance for which there was insufficient information to conclude whether the substance constitutes a danger to the environment, waste/used crankcase oils, 28 pp (August)
Environment Canada (2011) Follow up on the final decision on the assessment of releases of used crankcase oils to the environment, 13 pp (April)
Eriksson O, Carlsson Reich M, Frostell B, Björklund A, Assefa G, Sundqvist J-O, Granath J, Baky A, Thyselius L (2005) Municipal solid waste management from a systems perspective. J Clean Prod 13(3):241–252
European Union (EU) (2000), Directive 2000/53/EC of the European parliament and of the council of 18 September, 2000 on end-of-life vehicles, 15 pp
Ferrão P, Amaral J (2006a) Assessing the economics of auto recycling activities in relation to European union directive on end of life vehicles. Technol Forecast Soc Chang 73(3):277–289
Ferrão P, Amaral J (2006b) Design for recycling in the automobile industry: new approaches and new tools. J Eng Des 17(5):447–462
Ferrão P, Nazareth P, Amaral J (2006) Strategies for meeting eu end-of-life vehicle reuse/recovery targets. J Ind Ecol 10(4):77–93
Fletcher S (2011) Green parts. Claims Canada 5(4):46–47 (August/September)
Gente V, La Marca F, Lucci F, Massacci P, Pani E (2004) Cryo-comminution of plastic waste. Waste Manag (Oxford) 24(7):663–672
Gerrard J, Kandlikar M (2007) Is European end-of-life vehicle legislation living up to expectations? Assessing the impact of the ELV directive on 'green' innovation and vehicle recovery. J Clean Prod 15(1):17–27
Gesing AJ, Reno D, Grisier R, Dalton R, Wolanski R (1998) Non-ferrous recovery from auto shredder residue suing eddy current separators, minerals, metals and materials society (TMS). In: Proceedings of the 1998 TMS annual meeting, EPD congress, San Antonio, TX, 1998, pp 973–984 (February)
Haddad C (2008) Hybrid yards, some recyclers are offering customers twice as many options. Canadian Auto Recycl Mag 2(1):15–16
Hoeher M (2009a), Fluid drainage in modern day recycling enterprises—a survey on common practices. http://recyclerssource.com/PDFs/news/US%20report%20on%20Fluid%20Drainage%20in%20Modern%20Day%20Recycling%20Enterprises%20by%20Michael%20Hoeher.pdf . Accessed 28 Aug 2011, 15 pp (4 December 2009)
Hoeher M (2009b) A survey on fluid drainage practices of canadian automotive recyclers. http://recyclerssource.com/PDFs/news/Fluid%20Drainage%20Practices%20of%20Canadian%20Automotive%20Recyclers%20by%20Michael%20Hoeher.pdf. Accessed 28 Aug 2011, 15 pp (8 Dec 2009)
Hoeher M (2010) Fluid dynamics, a survey a survey on fluid drainage practices of canadian automotive recyclers. Canadian Auto Recycl Mag 4(1):74–77
Hopewell J, Dvorak R, Kosior E (2009) Plastics recycling: challenges and opportunities. Philos Trans R Soc B Biol Sci 364(1526):2115–2126
Institute of Scrap Recycling Industries (ISRI, Inc.) (1998) Clean Air Act title V applicability workbook. Prepared by Versar Inc., for ISRI, Washington, DC
Jekel LJ, Tam EKL (2007) Plastics waste processing: comminution size distribution and prediction. J Environ Eng 133(2):245–254

Jody BJ, Daniels EJ (2006) End-of-life vehicle recycling: the state of the art of resource recovery from shredder residue. argonne national laboratory, energy systems division. http://www.es.anl.gov/Energy_systems/CRADA_Team/publications/Recycling_Report_(print).pdf. Accessed 29 Aug 2011

Johnson MR, Wang MH (2002) Evaluation policies and automotive recovery options according to the European Union directive on end-of-life vehicles. J Automob Eng 216(part D):723–739

Kahhat R, Kim J, Xu M, Allenby B, Williams E, Zhang P (2008) Exploring e-waste management systems in the United States. Resour Conserv Recycl 52(7):955–964

Keoleian GA, Kar K, Marion MM, Bulkley JW (1997) Industrial ecology of the automobile: a life cycle perspective. Society of Automotive Engineers Inc, Warrendale, 159 pp

Kijchavengkul T, Auras R (2008) Compostability of polymers. Polym Int 57(6):793–804

Kim K-H, Joung H-T, Nam H, Seo Y-C, Hong JH, Yoo T-W, Lim B-S, Park J-H (2004) Management status of end-of-life vehicles and characteristics of automobile shredder residues in Korea. Waste Manag (Oxford) 24(6):533–540

Kim S, Dale BE, Drzal LT, Misra M (2008) Life cycle assessment of Kenaf fiber rein-forced biocomposite. J Biobased Mater Bioenergy 2(1):85–93

Loehr K, Melchiorre M (1996) Liberation of composite waste from manufactured products. Int J Miner Process 44–45:143–153

Michaels D (2007) Boeing and airbus compete to destroy what they built. The Wall Street J. Accessed 10 March 2011 (June 1)

Mirabile D, Pistelli MI, Marchesini M, Falciani R, Chiappelli L (2002) Thermal valorisation of automobile shredder residue: injection in blast furnace. Waste Manag (Oxford) 22(8):841–851

New Brunswick (2002) New Brunswick Regulation 2002-19, Used Oil Regulation—Clean Environment Act (O.C. 2002-95)

Ontario Automotive Recyclers Association (OARA) (2011) Effective and efficient end-of-life-vehicle environmental management in Ontario. http://oara.com/fileUploads/1311596363–OARA_ELVIS_White_Paper_FINAL.pdf. Accessed 12 Aug 2011, p 19

Ontario Ministry of the Environment (OMOE) (2009) From waste to worth: the role of waste diversion in the green economy. Minister's report on the Waste Diversion Act 2002 review, 38 pp (October)

Palmer J, Ghita OR, Savage L, Evans KE (2009) Successful closed-loop recycling of thermoset composites. Compos A Appl Sci Manuf 40(4):490–498

Pimenta S, Pinho ST (2011) Recycling carbon fibre reinforced polymers for structural applications: technology review and market outlook. Waste Manag (Oxford) 31(2):378–392

Puri P, Compston P, Pantano V (2009) Life cycle assessment of Australian automotive door skins. Int J Life Cycl Assess 14(5):420–428

Quebec (2008) Extended producer responsibility (EPR), current status, challenges and perspectives, développement durable, environnement et Parcs. http://www.mddep.gouv.qc.ca/matieres/valorisation/0803-REP_en.pdf, 145 pp (March)

Rao MD (2003) Recent applications of viscoelastic damping for noise control in automobiles and commercial airplanes. J Sound Vib 262(3):457–474

Recycling Council of Ontario (RCO) (1999). Management of end-of-life vehicles (ELVs) in Ontario, report, proceedings and draft recommendations of the RCO roles and responsibilities forum, 28 April 1999 (Sept)

Santini A, Herrmann C, Passarini F, Vassura I, Luger T, Morselli L (2010) Assessment of ecodesign potential in reaching new recycling targets. Resour Conserv Recycl 54(12): 1128–1134

Sasse F, Emig G (1998) Chemical recycling of polymer materials. Chem Eng Technol 21(10): 777–789

Sawyer-Beaulieu S (2009) Gate-to-gate life cycle inventory assessment of north american end-of-life vehicle management processes. Ph.D. dissertation, University of Windsor, Windsor, Ontario

Sawyer-Beaulieu S, Tam E (2006) Regulation of End-of-Life Vehicle (ELV) retirement in the US compared to Canada. Int J Environ Stud 63(4):473–486 (Aug)

Schluep M, Hagelueken C, Kuehr R, Magalini F, Maurer C, Meskers C, Mueller E, Wang F (2009) Recycling—from e-waste to re-sources. Sustainable innovation and technology transfer industrial sector studies, United Nations Environment Programme, Nairobi, Kenya. http://www.unep.org/PDF/PressReleases/E-Waste_publication_screen_FINALVERSION-sml.pdf. Accessed 29 Aug 2011
Schlummer M, Gruber L, Maurer A, Wolz G, van Eldik R (2007) Characterisation of polymer fractions from waste electrical and electronic equipment (WEEE) and implications for waste management. Chemosphere 67(9):1866–1876
Sinha V, Patel MR, Patel JV (2010) PET waste management by chemical recycling: a review. J Polym Environ 18(1):8–25
Spicer A, Wang MH, Zamudio-Ramirez P, Daniels L (1997) Disassembly modeling used to Assess automotive recycling opportunities. SAE Special Publications, 970416
Staudinger J, Keoleian GA (2001) Management of End-of-Life Vehicles (ELVs) in the US. Center for Sustainable Systems, University of Michigan, Ann Arbor, MI, report no. CSS01-01, 58 pp (March)
Strömberg E, Karlsson S (2009) The design of a test protocol to model the degradation of polyolefins during recycling and service life. J Appl Polym Sci 112(3):1835–1844
Struik LCE (1977) Physical aging in plastics and other glassy materials. Polym Eng Sci 17(3):165–173
Suddell BC (2007) The increasing trend of utilising biobased materials in automotive components. J Biobased Mater Bioenergy 1(3):454–460
Summerhill Impact (2011a) Retire your ride. http://www.retireyourride.ca/home/about-the-program.aspx. Accessed 29 Aug 2011
Summerhill Impact (2011b) Switch out program. http://www.switchout.ca/. Accessed 31 Aug 2011
The Alliance of Automobile Manufacturers (Auto Alliance) (2011) Recycling automobiles. http://www.autoalliance.org/index.cfm?objectid=B205C4D0-9054-11E0-A62C000C296BA163. Accessed 29 Aug 2011
The Society of Motor Manufacturers and Traders Limited (2011) Motor industry facts 2011. https://www.smmt.co.uk/shop/motor-industry-facts-2011-2/. Accessed 17 Aug 2011
Taylor B, Toto D (2006) Growth industry: the list of America's auto shredding plants has grown with the global metals boom. Recycl Today 44(10):50–106 (October)
Towle I (2007) The aircraft at end of life sector: a preliminary study. University of Oxford. http://users.ox.ac.uk/~pgrant/Airplane%20end%20of%20life.pdf. Accessed 29 Aug 2011
United States Environmental Protection Agency (USEPA) (1994) Managing used motor oil, EPA/625/R-94/010. Office of Research and Development, Washington, DC, 84 pp (Dec)
USEPA (2006) Memorandum of understanding to establish the national vehicle mercury switch recovery program, August 11, 2006. http://www.epa.gov/mercury/pdfs/switchMOU.pdf. Accessed 29 Aug 2011, 14 pp
United States (2010) On-site burning in space heaters, code of federal regulations, title 40, protection of the environment, part 279, standards for the management of used oil, §279.23, 2010 ed
Vidal R, Martínez P, Garraín D (2009) Life cycle assessment of composite materials made of recycled thermoplastics combines with rice husks and cotton linters. Int J Life Cycle Assess 14(1):73–82
Weiss MA, Heywood JB, Drake EM, Schafer A, AuYeung FF (2000) On the road in 2020: A life-cycle analysis of new automobile technologies. Energy Laboratory. Massachusetts Institute of Technology, Cambridge, MA, Report No. MIT EL 00-003, 160 pp (Oct)
Zia KM, Bhatti HN, Bhatti IA (2007) Methods for polyurethane and polyurethane composites, recycling and recovery: A review. React Funct Polym 67(8): 675-692

Implementation of ELV Directive in Poland, as an Example of Emerging Market Country

Paulina Golinska

Abstract In the European Union for over decade the End-Life-Vehicles (ELV) directive is shaping the automotive sector aftermarket approach. Poland is an example of the emerging market country which joint European Union, when ELV directive was already agreed. In Poland since 2006 efforts are made in order to organize the recovery network for ELV vehicles, but still a number of problems appear. This chapter focuses on the End-Life-Vehicles management. The state-of art is provided as well as the highlights for the improvement of current situation. The chapter presents the overview of the problems which appear by the configuration of the reverse logistics network. Author discusses the theoretical background, indicating main factors, which influence the scope and geographical distribution of the recovery network.

Keywords End-Life-Vehicles · Reverse logistics · Recovery network

1 Introduction

The life cycle approach is dominant in automotive industry. It is caused mainly by legal regulations which defined the level of vehicles recoverability. Even on the markets where the formal regulations don't exist regarding End-Life-Vehicles management, some actions are taken. Automakers try to achieve some additional profit from use of recycled materials and remanufactured auto parts or energy recovery.

The automotive industry is a sector where environmental impact must be taken into consideration in many ways. The beginning of life phase (BOL) focuses mainly on eco-design and making production processes to be less harmful for the

P. Golinska (✉)
Poznan University of Technology, Strzelecka 11 60-965 Poznan, Poland
e-mail: paulina.golinska@put.poznan.pl

P. Golinska (ed.), *Environmental Issues in Automotive Industry*,
EcoProduction, DOI: 10.1007/978-3-642-23837-6_11,

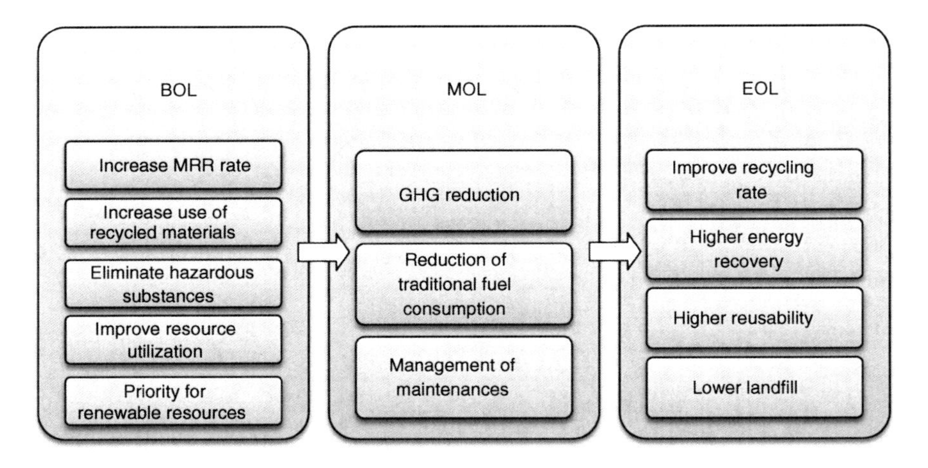

Fig. 1 Environmental focus in each life cycle phase

environment. Product itself must be optimized for the most eco-friendly usage within middle of life phase (MOL) and the least harmful in the end of life phase. The environmental focus in each life cycle phase is highlighted in Fig. 1.

In order to meet the legal regulation regarding End-of-life in the BOL phase decision must be made, which allow to apply engineering solutions that:

- steady increased material recovery rates (MRR) in EOL
- increase use of renewable resources and recycled materials—which build up demand for recycled materials,
- improvement of non-renewable resource utilization in manufacturing phase,
- reduction of hazardous substances like lead, mercury, cadmium, and hexavalent chromium.

The manufacturers make an effort to implement new innovations which let cars to pollute the air less. The reduction of car weight and fuel consumption is a good example of such actions. The main aspects which are taken in consideration are:

- fast growing number of cars and changes in consumption,
- high average age of cars used by consumers.

The number of vehicles including cars, light-, medium- and heavy-duty trucks and buses registered worldwide is rising. In 2010 for the first time the amount of vehicles on the roads has surpassed one billion (wardsauto.com/ar/world_vehicle_population_1108). The vehicles production in 2011 was over 80 million units (including cars and commercial vehicles) (http://www.oica.net, 2012). The average age of car in the European Union is over 8 years, for Poland is about 2–3 years more. The problem of old car is growing.

The problem of End-Life-Vehicles was addressed by the European Union by the elaboration of End-Life-Vehicles Directive (ELV 2000/53/WE) over a decade ago in 2002. It was a legal basis for the creation by all Member States of a unified system of waste management for old cars. Preparation of the directive was to establish a uniform system for recycling cars in all EU Member States, so as to avoid transport car wrecks to countries which have lower environmental requirements. Directive comprises the duties of users, producers and importers as well as establishments for their dismantling and recycling. The regulations contained therein relate not only to the vehicle as a whole but also for its components and materials. The Directives aimed to:

- prevent the creation of waste from vehicles,
- stimulate the re-use and recycling and other forms of recovery of end-of life vehicles and their components,
- reduce the quantity of waste for disposal,
- improve the environmental performance of the activities taken by all operators involved in the full life cycle of the vehicle and, in particular, directly involved in the processing of end-of life vehicles.

The ELV Directive requires recoverability and recyclability rates:

- from 1st January 2006 a minimum of 85 % of vehicles should be reused or recovered (including energy recovery) and at least 80 % must be reused or recycled,
- from 1st January 2015 a minimum of 95 % of vehicles should be reused or recovered (including energy recovery) and 85 % reused or recycled.

In order to reach the challenging goal of 95 % recovery target by 2015 some efficient material separation technologies for End-Life-Vehicles are promoted that allow the utilization for shredder residue and boosting the usage of recycled materials for some specific car components.

In Poland ELV directive was established in 2005. In the subsequent sections the main elements of the national recovery system will be presented. Author identifies the problems that still exist as well as main challenges for improvement of old vehicles reverse logistics activities in the future.

2 Poland as an Example of Emerging Market

Poland with the population of almost 37 million is one of the biggest countries in the European Union. At the same time it is the biggest emerging market in EU. At the end of the year 2011 in Poland were registered over 17,87 million passengers cars, about 53 % more than in 2003, just before the accession to the European Union. Including lorries and motorcycles and busses the vehicle fleet in Poland equals to over 23,85 millon units.

At the same time Poland still is the biggest importer of used car in the whole European Union. The import has rapidly grown after the accession to European Union in year 2004. The main reason for the ageing vehicle fleet in Poland after

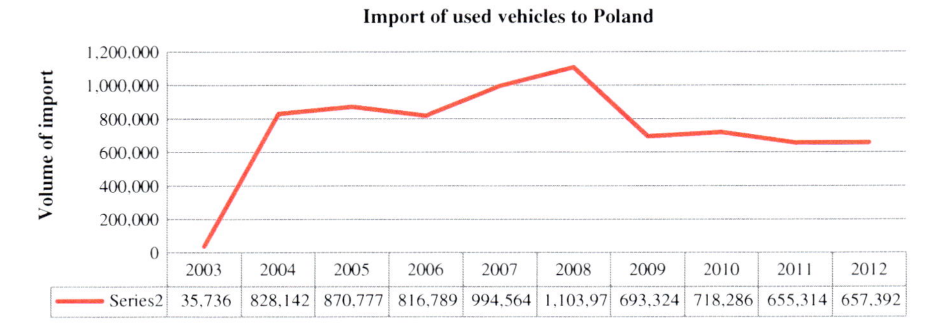

Fig. 2 Used cars' import to Poland

the accession to the EU was the abolishment of the Euro II standard. Before Poland's accession to the EU, when restrictions on import of older models were still in force, the segment of cars'aged five years and under had an over 18 % of share in total fleet. The abolition of the Euro II emission standards, allowed the freedom to purchase on the EU territory (the so-called acquisition within the community) cars which had been manufactured before 1997. Many cars after accidents or those aged over 10 years were imported. At present the yearly average import volume is almost three times higher than the new car production. The Fig. 2 presents the volume of used car import to Poland since accession to the EU in 2004.

The bigger problem is the fact that over 50 % of imported cars are over 10 years old. Average annual sales in the European Union account for 30 new vehicles per 1000 population. In Poland it is four times lower (Automotive Industry Report 2012). According to the official governmental database statistical vehicle registered in Poland is 15 years old. The average age of car which is disassembly in Poland is over 16 years. The experts however claim that the average age of vehicles is 11,2–12 years. The differences appear due problems with efficient updating of official the cars' register (the national cars' and drivers' database was finally established but the quality of the date inside is still doubtful). The age structure presents Figs. 3 and 4.

The problem with the assessment cars' age structure in Poland is mainly caused by inefficient central statistics system, so called CEPiK. Only cars which are officially dismantled are withdrawn from the statistics. Due to the "grey zone" (unofficial dismantling activities) a big number of cars exist only virtually in the documents. According CEPiK over 300,000 vehicles are declared off the road every year and their number is growing steadily. The standard for most of the European Union countries is renewal of 6 % of existing fleet per year. In Polish condition it means that over 1 million cars should be withdrawn annually.

In comparison with other EU countries the age structure is very bad. Many EU countries during the crisis have implemented the renewal scheme for the vehicle

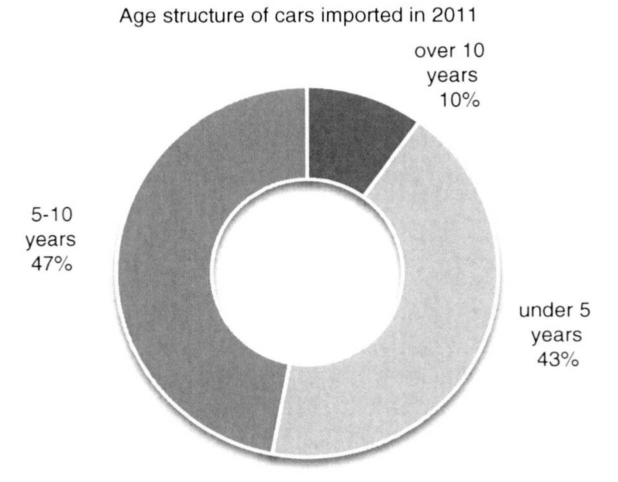

Fig. 3 Age structure of cars imported to Poland in 2011

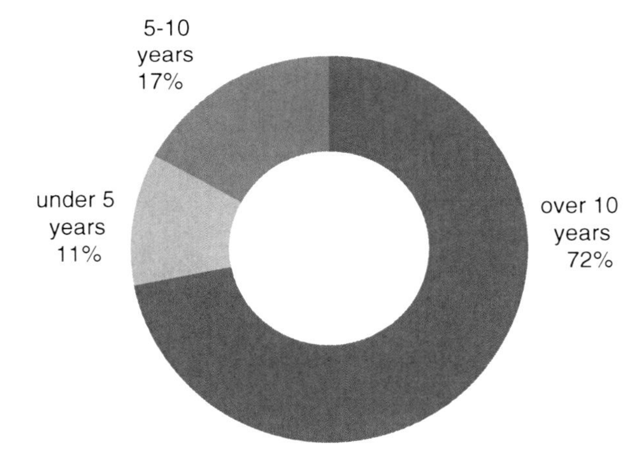

Fig. 4 Cars' age structure in Poland in 2011

fleet. The Governments of many European countries have developed an effective system in order to motivate drivers (financial incentives) to donate for scrap cars with more than 10 cars. The car owners received a bonus for new vehicles, which are more secure, more ecological and consume less fuel. The distance between Poland and the rest of the car industry in Europe is still deepening. In effect Polish market will be main place for End-Life-Vehicles disposal in the next decade.

The existing system is not efficient enough to handle such huge amounts of old vehicles. In the next subsections are described the main elements of Polish recovery system, as well as the main problems.

3 Recovery Network and Its Management

In the traditional supply chain information and material flows occur from one manufacturer or distributor to multiple locations, where the sale occurs. In the case of return logistics it is necessary to consolidate the material streams, coming from many distributed users. Recovery network is not a mirror image of the distribution network, in practice there are a number of differences in the organization of distribution channels and product recovery channels.

Companies seek to optimize processes in distribution networks by analyzing location distribution warehouses, reducing transport costs and minimizing inventory levels. This type of action is aimed at shortening the cycles of supply and to ensure customer service levels established at the lowest cost. Optimal recovery network organization requires similar approach. In order to optimize material flow in the recovery network, the following activities are necessary feedback:

- monitor type and the status of returning products (gatekeeping),
- shortening delivery cycles (called re-supply),
- consolidation of the warehouse processes for returned products (reverse consolidation centers).

Monitoring of the status of returning products (gatekeeping) aims to prevent unjustified/excessive material flows. This action allows reducing the cost of unnecessary transportation or reprocessing of products, which should be landfilled.

Shortening the cycles of secondary supply is intended to minimize the period of time that elapses from the moment of the decision to return the product and its reuse. Often, it is difficult to decide whether a product can be reused or whether it should be sent to the landfill. In the case of reverse logistics relatively infrequently used are advanced IT tools for decision making support. For this reason, a lot of information needs to be added manually, especially when it comes to the code that defines the reason for the return, the place of origin of the product and its technical condition.

Filling these gaps of information is particularly important when companies decide to create a consolidated sites collecting, sorting and processing of returns, where quick identification of the product model, its technical specification and the degree of wear allows making cost-effective decisions in a much shorter time. In the consolidated centers return processes are standardized, and therefore it is easier to identify and avoid potential mistakes. Choosing the best recovery network channel is critical to maximize the economical benefits.

Consolidation centers in recovery network should be separated from the existing distribution centers within the supply chain. This is due to the fact that combining in one location primary and secondary flow in the supply chain always causes situation where the primary flows will have precedence over reverse ones. This approach has had a negative impact on the sustainability of lead times in reverse logistics networks (re-supply lead times).

3.1 Recovery Network Organization

The configuration of a supply chain is the particular arrangement or permutation of the supply network's main elements, including the network structure of the various operations within the supply network and their integrating mechanisms, the flow of materials and information between and within key unit operations, the role, inter-relationships, governance between network partners, and the value structure of the product or service delivered (Srai and Gregory 2009). The definition is also suitable for configuration of recovery network.

The configuration process is an arrangement of parts or elements that gives the whole its inherent form. Parties involved in the cooperation have their own resources, capabilities, tasks, and objectives so there are difficulties in coordination of the constant flows of information, materials, and funds across multiple functional areas both within and between chain members (Golinska 2009).

Recovery network configuration requires defining scope and geographical aspects. The scope definition requires the designation which participants/intermediaries should be included. Recovery network is very rarely including direct channels fully managed by the manufacturer. More often companies create multilevel indirect channels through distributors, retailers, core brokers, recycling companies, etc.

Configuration in terms of the geographical location of the sites means the designation of individual points of the consolidation of the material streams. Impact on recovery network organization has also type of the reverse logistics strategy adopted to manage returns. Network configuration is different, depending on the chosen strategy, in case of:

- cost-oriented strategy—the aim is to achieve the lowest unit cost of product collection (including warehousing, transport and sorting), reprocessing and redistribution,
- time-oriented strategy—the aim is to achieve the shortest lead- time for reverse logistics processes (e.g. collection, reprocessing, redistribution).

Time-oriented strategy is more suitable for innovative products that quickly lose their value (for example, electronic equipment, computers, etc.). While cost-oriented strategy better works for products, which value is reduced in slow pace. In a situation cost-oriented strategy is used, it is recommended to create a centralized recovery network. This approach allows achieving economies of scale, and thus minimizing unit cost of reverse logistics processes, including transportation. In a decentralized recovery network most operations associated with the gatekeeping are carried out at the points where the customers make returns of product.

The previous works on recovery network configuration had taken in consideration costs of investments or operational costs in order to find the fixed geographical location of new facilities/points for recovery and product collection [eg. Beamon and Fernandes (2004)]. In case of ELV Directive the cost element of configuration is important but not dominant. Some legal restrictions must be also taken in consideration. In the next section the ELV recovery network is described for Poland.

3.2 ELV Recovery Network in Poland

In Poland the directive ELV was implemented by introduction in the 20 January 2005 the law act recycling of End-Life-Vehicles (Dz.U, 2005 No. 25 item 202, with further amendments.). The Act specifies the obligations of manufacturers and importers of cars during the recycling process, including car delivery to the dismantling stations detailed handouts from the scope of the manual dismantling of vehicles and the location of hazardous materials in the car. In addition, the car manufacturers and importers are required to pay all or a significant part of the costs of providing vehicles with zero or negative market value to dismantling stations, without expense to the last user or owner of the vehicle.

The recovery network must follow the legal requirements regarding:

- distance between particular dismantling facilities,
- characteristics of the minimal technical resources and employees competences.
- The main participants of recover network are:
- car manufactures/or car importers
- dismantlers
- recyclers
- remanufacturers
- car users.

From point of view of recovery network configuration the most important requirements of the Polish regulations about ELV were:

- producers and importers of more than 1000 units per year needed to ensure that the owner was able to left the End-Life-Vehicle at the point of collection or dismantling, situated at a distance of not more than 50 km in a straight line from the place of owner's residence.
- if the above mentioned requirement was not satisfied, that manufacturer or importer needed to pay the fee of 500 PLN (approx. 125 euro) per every vehicle sold.

According to the data collected by the Polish Car Recycling Forum (a sector's association) the number of dismantling station growing on the stable rate (http://fors.pl/pliki/ilosc_prowadzacych.pdf, 2012). The detailed numbers are provided in Fig. 5.

In 2011 the Polish ELV regulation was amended and the fee for producers and importers is not valid any more. The reason for such change was the fact that Polish Government assumed that the basic network of dismantling stations had been already provided. At the end of year 2011 there were about 740 dismantling stations and 118 collection points. In comparison, at the beginning of recovery network creation period (at the end of 2005) the number of dismantling station was 360 and there were about 50 collection points.

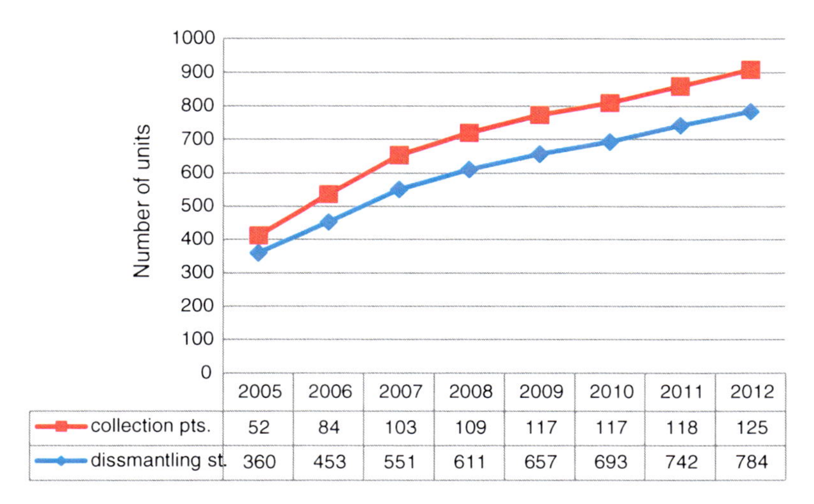

	2005	2006	2007	2008	2009	2010	2011	2012
collection pts.	52	84	103	109	117	117	118	125
dissmantling st.	360	453	551	611	657	693	742	784

Fig. 5 Number of collection points and dismantling station

The dismantling stations are the only officially approved entities, which are allowed to process the End-Life-Vehicles. In order to get the permission to open the dismantling station, the owner must provide the defined by government infrastructure and skilled personnel. The dismantling stations are under constant monitoring and are required to achieve the goals:

- reusability and/or recyclability of at least 80 %, and reusability and/or recoverability of at least 85 % by weight, if measured against the international standard ISO 22620 (for transition period till 31st December 2014)
- reusability and/or recyclability of at least 85 %, and reusability and/or recoverability of at least 95 % by weight, if measured against the international standard ISO 22620 (starting form 1st January 2015).

The only exemption from these rules are the cars which were manufactured before 1st January 1980. For them the rates are: reusability and/or recoverability of at least 75 and 70 % recyclability.

From these ELVs, dismantling companies first remove the oils, engines, transmissions, tires, batteries, catalytic converters, and other parts, which are then recycled or reused. Shredding companies then sort out the ferrous and non-ferrous metals, resins and other materials. While the ferrous and non-ferrous metals are recycled, the remaining material is collected by manufacturers and recycled/recovered or processed appropriately.

A big problem for Polish recovery network is the fact that in years 2006–2011 there were financial subsidies for small dismantling stations with capacity below 500 cars per year. This situation had discouraged the dismantling stations' owners to invest in well-equipped medium/high capacity facilities. In Polish conditions dismantling station that proceed per annum over 1000 End-Life-Vehicles is perceived as a big one.

Another problem regarding the recovery network is the existence of "grey zone". A big number of not officially approved dismantling stations exist. These stations don't invest in the infrastructure required by legal acts and are not reported in any official statistics. Every final owner of the vehicle may handover his or her old vehicle free-of-charge, only when it is delivered as complete and uniquely identifiable (VIN number or chassis number, bodywork or frame). If the vehicle is incomplete and the vehicle owner refuses payment, the operator may refuse its acceptance. According to the ELV Act for the complete vehicle shall be deemed to be a vehicle that contains all the crucial elements and its mass is not less than 90 % of the mass of the vehicle. The owners of incomplete vehicles can be charged up to 10PLN (about 2.5 euro) for every kilogram of missing mass of the car. Because of this, often the owners of such vehicles, often prefer to deliver the incomplete ELV to small mechanical workshops, which are not authorized for dismantling. The ELVs are illegally dismantled and used as a source of spare parts, waste that has no value is abandoned on fields or in forests. The cost of dismantling operations in illegal stations is much lower than in the approved dismantling stations. The dismantling operations there are done without fulfillment of legal regulations and are dangerous for natural environment and workers. The size of "grey zone" can only be estimated. As mentioned before the typical cars' renewal scheme for markets like Polish is 6 % per year. It means that about 6 % of old cars should be deregistered every year. In Poland according to the official statistics only about 265 000 t cars in 2010 and 342 352 in 2011 were withdrawn. It means that only 1/3 of cars are dismantled in the official dismantling stations (FORS 2010). Most of the illegal dismantled cars are sold by the owners as spare parts to the secondary market. The rest of car is probably landfilled or left illegally at the parking lots or forests. There are no financial incentives for car's owners for delivering the vehicle to the officially approved dismantling station. A comparison of the size of the fleet and the small number of vehicles handed over to legal dismantling facilities clearly indicates the Polish vehicle recovery system does not operate correctly.

The situation is more and more dangerous for the natural environment. It needs to be highlighted that the problem will be growing because of the age structure of cars' market.

4 The Challenges for the Future

Recovery network configuration is still a major challenge, regarding the elimination of illegal dismantling stations and providing for those legal an expected volume of the ELV. Restrictive legal provisions necessitate continuous improvement commitment levels of end of life products, as well as their efficient reprocessing. The main challenges for reverse logistics in the management of End-Life-Vehicles are to ensure an effective collection and organization of the economically profitable dismantling only in legally established enterprises with appropriate permissions. In order to obtain the required statutory levels of recovery, it is also necessary to create consolidation centers within recovery network

The ELV cars can be subject of different reuse options as:

- recycling,
- reuse as it is,
- remanufacturing.

In the automotive industry demand for recycled materials for many years will be lower than the potential supply. In order to utilize the End-Life-Vehicles at optimum level, there should be taken in considerations other ways of recovery of End-Life-Vehicles. Emphasis should be placed on remanufacturing and reuse-as-it-is of parts for used vehicles repairs. This forces the further development of the recovery network for End-Life-Vehicles. It is necessary to develop more advanced tools to monitor the condition of the parts in the dismantling stations, so as to ensure a stream of material suitable for the remanufacturing sites. In Polish conditions reuse "as it is" is very popular option. Many legally functioning dismantling stations try to sell the used auto-parts in order to achieve additional profit. The importance of remanufacturing is also growing, mainly due to the lower than the EU average work costs.

It is assumed that in the next decade Poland is going to be main automotive remanufacturing site. Growing the amount of remanufacturing activities requires the improvement of ELV returns management. The main problems in materials management for remanufacturing purpose can be defined as (Golinska 2009):

- the uncertain timing and quantity of returns,
- the uncertainty in materials recovered from return items,
- the present Polish reverse logistics network configuration,
- the problems of stochastic routings for materials for remanufacturing operations and highly variable processing times.

Most of the above mentioned characteristics result from lack of appropriate information on material flows and its forecasting. In Poland none IT tools are implemented on the national or regional level to forecast and monitor the End-Life-Vehicles returns. In order to reach stated in ELV directive goals a improvement is needed. Figure 6 present reference model for Polish ELV management in year 2015.

Taking in consideration all mentioned above problems the main challenges for Polish ELV recovery network and reverse logistics can be defined as:

- improvement of gatekeeping for car import,
- improvement of the collection of used cars at the officially improved dismantling stations and collection points,
- improvement of information management between recovery network participants,
- development of remanufacturing practices (including better materials management for remanufacturing).

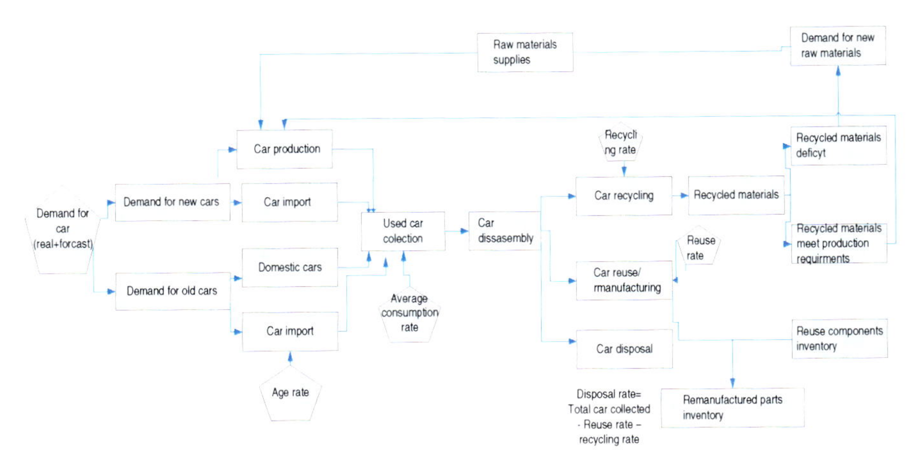

Fig. 6 Reference model for Polish ELV management in year 2015

The forthcoming years are going to be very challenging, especially regarding the need to improve the size/volume of dismantling stations throughout rather than their number. The better monitoring of the market is needed in order to eliminate existing dismantling "grey zone". Another challenge is still the full implementation of 2015 recovery/reusability and recyclability targets.

References

Automotive Industry Report (2012) Polish Automotive Association, Download from http://www.pzpm.org.pl/Rynek-motoryzacyjny/Raporty. Accessed 15 March 2012

Beamon B, Fernandes C (2004) Supply-chain network configuration for product recovery. Prod Plan Control 15(3):270–281

Golinska PI (2009) The concept of an agent-based system for planning of closed loop supplies in manufacturing system. In: Omatu S et al (eds) IWANN 2009 Part II. LCNS, vol 5518. Springer-Verlag, Berlin, pp 346–349

Golinska P, Kawa A (2011) Remanufacturing in automotive industry: challenges and limitations. J Ind Eng Manage 4(3):453–466 (Special Issue)

Guide VDR Jr (2000) Production planning and control for remanufacturing. In J Oper Manage 18:467–483

Lund R (1983) Remanufacturing: United States experience for developing nations. The World Bank, Washington, DC

FORS (2010) Regulacje prawne a praktyka i rzeczywistość, Eurorecykler II 1/2010

Srai JS, Gregory MA (2009) Supply network configuration perspective on international supply chain development. Inter J Oper Prod Manage 26(5):386-411

Steinhilper R (1998) Remanufacturing the ultimate form of recycling. Frauenhoffer IRB Verlag, Stuttgart

http://fors.pl/pliki/ilosc_prowadzacych.pdf (2012) Approach 15 March 2012

www.apra-europe.org (2012) Approach 15 March 2012

www.oica.net (2012) Approach 15 March 2012
www.remanufacturing.org.uk (2012) Accessed 15 Sept 2009
wardsauto.com/ar/world_vehicle_population_110815 (2012) Approach 15 March 2012
Nineth sustainability report-the UK automotive data (2007) www.smmt.co.uk Download 15 Sept 2009

Printed by Publishers' Graphics LLC
DBT140313.15.17.198